AI Transformation of Cities for Climate Solutions

Kwi-Gon Kim
Editor

AI Transformation of Cities for Climate Solutions

Springer

Editor
Kwi-Gon Kim
Seoul National University
Seoul, Korea (Republic of)

ISBN 978-3-032-21939-8 ISBN 978-3-032-21940-4 (eBook)
https://doi.org/10.1007/978-3-032-21940-4

This Springer imprint is published by the registered company Springer Nature Switzerland AG
The registered company address is: Gewerbestrasse 11, 6330 Cham, Switzerland

The original version of the book has been revised. A correction to this book can be found at https://doi.org/10.1007/978-3-032-21940-4_18

Introduction

COP 30 made meaningful strides in strengthening climate action at city, state and regional levels (World Resources Institute, Views Online, *COP30 Ends with Mixed Results*, 23 November 2025) and made one message clear: Cities are now at the center of global climate action (OpenEarth 2025) so it is time to put AI cities on the global stage.

A city is a living laboratory of artificial intelligence (AI) technologies. It is a laboratory where AI applications for city climate planning models are carried out and experiences are tested. AI could reset and reshape the future of our cities. AI transformation is part of the widespread green transition field. AI technologies have the potential to provide solutions that promote more efficient, time-saving urban planning and development processes, reduce carbon footprint, improve services for residents and increase the overall quality of life in cities.

Cities are a major cause of climate shocks. It is important to note that AI is based on quantitative data and computational logic—a machine learning process. A question here is: How can planning and development processes be combined with machine learning to provide solutions for climate shocks? And is this for or against 'sustainability'?

Convergence of Theory and Technology Towards New Sustainability—AI Nexus for City Climate Planning

Artificial intelligence, big data and climate change are the key areas to solve the challenges facing our planet in the next ten years. United Nations Sustainable Development Goals (UN SDGs) suggest criteria and indicators of sustainable development. These indicators can be used for the integrated consideration of environmental, social and economic sustainability of city climate planning and decision making in the urban context.

Smart and sustainable cities can be designed, operated and optimized, based on sustainability indicators for smarter climate solutions with AI, to find the right balance between automation and human judgment. AI transformation of cities and application of AI in climate city planning, within the broad concept of optimum urban sustainability, are based on sustainability cost and benefit, in order to leave a positive footprint. With AI, these themes could represent more than an incremental improvement of the traditional approach.

AI Agents and Platforms to Shape the Future of Cities with Climate Solutions

AI agents in urban planning mean much more than simple chatbots. Even though the climate-responsive urban planning and development processes are technically complex, two significant components in designing the process are prediction (or forecasting) and optimization for decision making, both of which can be improved by using AI platforms with tangible impacts. The use of AI platforms can aid city climate planners, urban planners and developers in modeling future change based on a sustainability-indicator approach. This involves optimizing climate-smart and sustainable eco-city solutions using AI algorithms, or intelligent algorithms, within the boundaries of data availability.

AI application in a systematic and holistic manner remains perhaps the single most underdeveloped area of urban planning and policymaking. While this situation is quickly changing with the emergence of new research and development in smart and sustainable eco-city solutions, city planners, managers and policy-makers still attempt to use AI in limited areas and with very little knowledge or experience of the life cycle value chain over the whole planning and policy-making process. Most planning and development systems do not yet explicitly address issues of AI to move climate-responsive development forwards. There is a need for reorienting research and development towards a new AI-based planning and development process that incorporates AI activities and tools for decision making, including the entire cycle of planning, construction, operation, monitoring and evaluation.

Objectives and Main Themes

AI is a new power for creativity. AI servers (Google Cloud or Alibaba Cloud) are finding their way into all aspects of city operations, including public safety, planning, transportation, and citizen service. This book considers theories, methods and potential applications of AI in climate-responsive urban planning and decision making—a combination of AI and city climate planning. This book intends to connect climate issues, urban planning and AI business and to invent new frameworks and methodologies from synthesis science perspectives. It explores how to build city climate optimism with support from AI in the age of climate crisis. Specific key objectives of this book are:

- To develop a theory of climate-smart and sustainable urbanism and synthesis science that create images and visions of tomorrow's cities with exponentially growing AI techniques.
- To explore actionable frameworks, methodologies and tools for incorporating AI technologies into city climate planning in the context of broader urban planning.
- To showcase successful applications that have strengthened climate action in urban areas.
- To find ways to foster networking, partnerships and cooperative work between urban planners, city climate planners, data and AI scientists, local governments, organizations and businesses.

This book is dedicated to advancing AI and other digital technologies, promoting sustainable transformation in related fields, and providing a platform for scholars, researchers, and practitioners worldwide to share and discuss innovative ideas and findings in terms of industry 5.0 or business 5.0.

It proposes that cities should develop the 'AI-Based Sustainability Navigator', an AI assistant designed to help cities navigate the complexities and nexus of themes such as sustainability, net-zero emission, biodiversity, and resilience. It provides AI technical guidance and information on challenges, solutions and financial options, including promotion of AI startups and business opportunities.

The main themes involve the logical steps of describing AI urban philosophy; AI frameworks, methodologies and tools for a new climate-smart urbanism approach; AI transition strategies and action plans; AI startups and case studies for networking; and prospects for AI city climate planning. Each stage is illustrated by examples of current AI activities and tools that can act as AI models to shape future practice and help make decisions.

The insertion of AI-based climate sustainability into the city could introduce further dimensions in urban planning in terms of innovative urban structures and functions. In particular, the newly emerging urbanism tries to unlock the power of AI with the next Generative AI (e.g. ChatGPT and Gemini). It enables us to connect AI climate action at all levels, using AI to link local action with national climate plans and global sustainable development goals.

The idea of the AI climate-smart sustainable eco-city could result in the establishment of an extensive urban digital network with data and AI platforms that could actually support climate and biodiversity solutions for sustainability, co-prosperity and peace beyond boundaries. It may become the core element in the new ideology of urban planning and development in a progressive AI era.

Structure of the Book

As the table of contents shows, the book is divided into seven parts.

Part I is concerned with how the new disruptive technologies have activated smart climate solutions, giving rise to the increasingly prevalent belief that climate-smart and sustainable eco-cities can help in achieving urban sustainability for climate solutions. This leads on to a discussion of new climate-smart, sustainable eco-urbanism.

Part II provides the methodology which theorizes AI transformation of cities and devises strategies and action plans in the interests of new urbanism and other positive values. It is based on a review of AI applications for climate solutions in terms of status and trends.

Part III discusses what is happening to contemporary cities in the context of the new AI urbanism that is actually changing urban structure and function. It describes the current status and trends in AI in cities.

Part IV addresses recent evolutions and practices in terms of AI technologies, startups and business.

Part V attempts, against the backdrop illustrated above, to show how to reorient urban planning and development in the form of an AI-based sustainability navigator or guide. It involves implications and suggestions for the future use of AI platforms or AI agents in cities, including designing future AI programs and projects (e.g. prediction of climate impact with AI, optimization of climate action for decision making with AI). The AI applications for urban policy, planning, program and project (the '4Ps') can be downscaled and divided into several levels of cities such as city, district, neighborhood, block and site.

Part VI addresses the networking initiative on AI applications for climate action by using a developed AI climate city planning model.

Part VII explores the future of AI transformation of cities with some inconclusive conclusions for broader AI technology research and applications of AI.

Future Prospects

This book is a synthesis of the author's five publications from Springer Nature that focused largely on the role of AI in cities.

The concept of AI transformation of cities could improve the quality of urban planning and development processes by integrating innovative AI technologies and knowledge. However, there is still a lack of general consensus about the goals and benefits of this initiative. In this context, this is a seminal volume and should be read with the understanding that it must be supplemented by additional research and development on this topic, with more global applications of AI agents for urban project planning and execution in a transformative era.

This work can be considered as a contribution to extend the body of knowledge in a crucial research stream. It may also support AI scientists and experts in designing future AI programs and projects in the urban context. The demonstration case studies may be the first applications of an innovative multi-agent as an integration platform solution, all from single platform. It could be a new brand of AI-driven cities. The brand value shall be great!

Our proposal of the Song Dam Lake is a unique case study, and it has been developed with local partners for implementation. As this book deals with participative urbanism, the citizen participation comes at an earlier stage, that is in the planning stage, and throughout the other stages. We also include the concept of citizen science, as well as local and indigenous knowledge, to complement the AI tools. With local participation we can gain site specific socio-ecological knowledge of the site, and build stewardship. The networking case studies propose Port Moresby City, Papua New Guinea, Kampala, Uganda, and the Caribbean region representing each continent.

Even though this book focuses on climate data acquisition and solutions, it also covers advanced topics associated with how AI technology works. Therefore, it can be used by professionals working on climate-responsive urban planning and design, city scientists, model-builders and users, ecologists, environmental managers, sustainability scientists, climate-impact scientists, synthesis scientists, data and AI scientists, engineering and computer science researchers and policy decision makers. It is hoped that this book will help us turn climate momentum into breakthroughs.

Reference

World Resources Institute (2025) 'COP30 Ends with Mixed Results', Views Online, 23 November
OpenEarth (2025) November and December 25, Newsletter

Contents

Part I
Basics of AI Transformation: Combining AI and Cities for Sustainability and Prosperity

Chapter 1
Potential of AI for Promoting Smart and Sustainable Eco-City Solutions to Climate Change

Sustainability–AI Nexus for AI City Climate Planning

Kwi-Gon Kim

Abstract AI is a new smart power which combines software and hardware power. Inputs are required from the AI discipline to ensure that a complete data analysis, prediction and optimization for urban decision making is accomplished. This chapter presents some prerequisites for proper usage of AI in terms of the role of AI in science and technology, requirements for a climate-responsive urban planning and development process, and green and sustainable AI technologies.

Keywords AI era · Optimum sustainability · AI sustainability · AI sustainability cost-benefit analysis · City climate optimism · Sustainable transformation · Age of climate crisis · Sustainability analytics · AI optimism · Scaling solutions · AI creative disruption

1.1 Role of AI in Science and Technology for Creative Disruption: Connecting AI with Cities for Sustainability

The emerging city infrastructures and spatial arrangements of the digital era profoundly affect all aspects of cities. While digital technologies have provided numerous benefits for cities, urban planners often face problems with the role of AI in the urban context because of the lack of AI technological discussion within urban academia.

Although urban concerns are crucial in discussions about AI, matters related to AI in urban planning and development have been largely missing from both professional and academic debates. The potential roles of urban policy-makers and planners in AI

K.-G. Kim (✉)
Seoul National University, Seoul, South Korea
e-mail: kwigon@snu.ac.kr

K.-G. Kim (ed.), *AI Transformation of Cities for Climate Solutions*,
https://doi.org/10.1007/978-3-032-21940-4_1

new technologies in cities have also been overlooked at the local level. This neglect is problematic because cities can be a living laboratory of urban experimentation with AI and cloud-based technologies (e.g. AI servers and cloud services) leading to a well-planned city. We need to understand the relations between cities and AI, and the relations between human intelligence and machine intelligence, considering the 'local' and the 'urban' in an integrated 'loci' manner. AI is a type of intelligence, demonstrated by a machine, that mimics a human's natural intelligence. In this context, conventional urban planning and development approaches should be completely changed to include more innovative AI agents.

1.2 AI Technology in Climate-Responsive Urban Planning and Development Processes: Reducing CO_2 Emissions with AI Technology

Even though a comprehensive report on current status and trends in the AI transformation of cities for climate solutions is not available, literature reviews reveal that numerous AI technologies have been used to offer innovative climate change mitigation, adaptation and cross-cutting of climate action measures. Table 1.1 shows a summary of AI applications for city climate planning.

In Table 1.1, we can see that AI helps to incorporate both citizen science data and the concept of engaging the public with the data being collected. The opportunities to leverage AI to deepen the public's commitment to the activities and to support maintenance and investment in the area will be exciting and novel.

As illustrated in Table 1.1, AI is increasingly being used in city planning and policymaking. By simulating population growth, land use and infrastructure stress points, planners can evaluate the long-term consequences of zoning decisions, construction projects or environmental policies.

AI-based urban planning could make a significant contribution to climate-responsive sustainable development, land-use patterns, urban forms, urban infrastructure and the structure of the urban environment. It is important to incorporate powerful AI practices with urban planning for the age of climate change. There are many technically isolated examples of AI good practices that could be systematically applied in city climate planning through a systematic and holistic approach. Climate infrastructure refers to the systems that enable city planners to use AI technologies. However, AI-powered lifestyle changes and local authority transformations present challenges in reducing urban greenhouse gas (GHG) emissions.

Electricity consumption by AI data centers is rapidly increasing due to increasing use of GenAI. In these circumstances, green and sustainable AI is becoming an alternate way of protecting the environment and reducing CO_2 emissions. Sustainable AI technologies need to be developed, both for maintaining AI performance and for reducing the impact of energy consumption on the environment, by adopting effective learning methods and environment-friendly AI systems. Here are some examples:

Table 1.1 Summary of AI applications for city climate planning: climate AI frontiers

Climate action	Categories	AI Applications	Tools
Mitigation	• Climate modeling and prediction	• AI enhances traditional climate models by handling vast datasets and detecting complex patterns	• Improved climate forecasting using deep learning models (e.g. temperature, rainfall, extreme weather)
	• Remote sensing and data analytics	• Land use and vegetation monitoring: tracks changes in green cover and urban sprawl	• Downscaled global models for regional/ local predictions
	• Energy and emissions management	• Carbon emission tracking: AI models estimate emissions by sector (transport, buildings, industry) using sensor and usage data • Smart grid optimization: AI forecasts energy demand and supply, helping cities balance renewable energy integration • Energy efficiency planning: AI identifies buildings with high energy consumption and suggests retrofitting strategies	• Copernicus data store • Planet LAB + AI analytics • NASA's EarthData with ML plug-ins • Carbon Aware SDK (by Green Software Foundation) • EnergyStar Portfolio Manager (with AI driven-analytics) • Google DeepMind's energy optimization for buildings
	• Nature-based solutions (NBS)	• Identification of optimal locations for green roofs, urban forests, and water-absorbing surfaces (rain gardens)	• UrbanFootprint (planning platform using AI)
	• Transportation and mobility planning	• Traffic flow optimization • Public transport planning • EV infrastructure planning	• CitySim or ENVI-met (AI-enhanced urban simulation tools) • Remix (transport planning tool) • PTV Visum (with AI-based traffic simulations)

(continued)

Table 1.1 (continued)

Climate action	Categories	AI Applications	Tools
Adaptation	• Climate finance and risk assessment	• AI assists with climate risk modeling and sustainable investments	• Assessing physical and transition risks for businesses • AI-powered ESG analysis • Insurance modeling for climate-related disasters • AI tokenization
	• Climate risk assessment and prediction	• Urban heat mapping • Flood risk modeling • Sea level rise simulations • Air quality forecasting • Infrastructure vulnerability analysis • Early warning systems	• Google's Environmental insights • Explorer • Microsoft's AI for Earth • IBM's Green Horizons
	• Disaster detection and response	• Real-time analysis of wildfires, floods or storms using satellite imagery	• Google DeepMind's GraphCast
Cross-cutting climate action measures	• Citizen participation and policy modeling	• AI chatbots for citizen feedback: collects input on climate-related issues (e.g. flooding, heat complaints) • Sentiment analysis on climate policies: • Understand public opinion using AI on social media or surveys • Participatory planning platforms: AI helps model outcomes based on public input	• ZenCity (AI-based citizen sentimental platform) • Decidim (open-source participatory planning, integrated with AI tools) • Perflexity (a free AI-powered answer engine that provides real-time answers to any question)

(continued)

Table 1.1 (continued)

Climate action	Categories	AI Applications	Tools
	• Policy and behavioral change	• AI can inform and influence decisions by policymakers and the public	• Climate scenario simulations for policymakers • Natural language processing (NLP) for analyzing climate legislation or misinformation • AI chatbots and apps that help users track and reduce their carbon footprint

- Use of cooling systems and renewable energy for AI servers.
- Automation of AI agent model development, distribution and management for environmental and climate monitoring.
- Optimization of data and information communication services.
- Use of AI technologies to create environmental value.

There are critical challenges in the environmental sustainability of climate-stressed cities. We need to facilitate innovative development of AI to reshape the way we design, manage and experience cities to consider the future sustainability of cities. It is important to note that the integration of AI into urban planning is opening new frontiers. Noyman's presentation delved into the transformative impact of agent-based modeling, generative AI (ChatGPT, Google Gemini, etc.), and large language models (LLMs) in urban environments, offering a glimpse into the future of urban planning. He suggests that AI agents could serve as powerful tools for evaluating urban policies (Noyman 2025).

Concerns have been raised on social media about biases in AI systems. Even though AI agents have moved beyond ChatGPT and currently available technology, investors commit large sums to the development of AI hubs mainly for Open AI ChatGPT, Google Gemini, Chinese DeepSeek and Kimmy 2, super data centers, unmanned mobility and robots.

There are other areas we need to explore in AI applications. According to recent research conducted in Australia, the AI market for urban management should not be ignored (Australian Trade and Investment Commission 2025). Urban AI systems science will unlock cities as major data hubs for emerging sovereign AI data.

Cities could establish small-scale AI pilot programs to reduce the environmental hazards (e.g. energy consumption, heat) associated with large-scale data center installations. Experimental AI techniques allow urban planners and policymakers to detect possible problems, collect useful data, and fine-tune technologies for cloud data storage and analytics, focusing on optimum land use, energy, buildings, water, waste, transportation, green space, climate and flooding. These AI programs can be based on real-time performance and community feedback, and decision-supporting systems.

Here are some suggestions for integrating AI systems into cities to help them work smarter, that is, more intelligently, efficiently and quickly.

First, AI urban systems should be demand-oriented: AI-driven applications in cities could offer increased proximity to personalized-service AI that enables data scientists to collect real-time demand/consumption-based data.

Second, new AI applications should be value-oriented: sustainability is a global common value and principle on which all AI technology should be based. To establish this common value and principle, local sustainability indicators should be used for AI architecture. Currently sustainability indicators are not utilized to create AI models and algorithms for predictive analytics and the optimization of programs and projects.

Third, AI innovations need new ways of thinking and options for the vision, policies and technologies, providing comprehensive perspectives on the interplay between technical advancements and social dynamics within urban environments. The needs should be implemented through a whole-society approach rather than limited to only science and technology and ICT.

It is time to see the AI potential in urban planning and management systems move from theory to action.

1.3 Data Availability for AI Agents, and AI Data Centers and Their Climate Impact

AI agents follow data availability. Data available from the cloud refers to information stored on remote servers, accessible over the internet, and managed by cloud providers. These data can include a wide range of information, from files and databases to applications and entire computing environments. Cloud storage, a key component of cloud computing, allows city climate planners to store data on virtual machines within data centers, offering accessibility, redundancy and scalability.

An AI data center is a specialized computing facility designed to handle the heavy computational demands of artificial intelligence workloads. These centers house advanced hardware (e.g. graphic processing units or GPUs), high-speed networking and efficient cooling systems to support AI training, deployment and operation.

As AI applications become more prevalent in cities, the demand for AI-specific data center capacity is increasing rapidly. AI data centers form the backbone of cloud computing platforms, providing the infrastructure for delivering AI-powered services to users. Table 1.2 shows the status of AI data centers.

There will be increased power consumption and heat release by the operation of AI data centers with cloud systems like a data factory (e.g. an AI internal path) that are contained in Table 1.2. AI should provide the green transition to net-zero emissions (in particular, Scope 3 emissions and indirect carbon footprint).

In training large AI models, the engines behind the tools that have now become a part of daily processes require immense computational power. Recent studies suggest

Table 1.2 Status of AI data centers

Global AI business	Location	Number of graphic processing units
Naver	Korea	2,000
Private-Government Joint Venture	Korea	10,000 (target in 2025)
OpenAI, SoftBank	Japan	100,000
xAI	USA	200,000
META	USA	1,300,000 (target in 2025)
OpenAI	USA	400,000
Google	USA	Not available
Alibaba	China	Not available

Source Chosun Daily News, 16 June 2025 (modified by the author)

that developing a single large model can produce emissions comparable to those of several cars over their entire lifespans.

Take ChatGPT, for example. It was estimated that training GPT-3 emitted over 550 metric tonnes of CO_2, roughly the same as driving a car around the world more than 100 times. Its successor, GPT-4, is widely believed to have a significantly larger footprint, likely running into several thousand metric tonnes of CO_2, though precise figures remain undisclosed. And that's just the start. Every report generated, dataset analyzed or prompt answered by these models draws additional power from an already stretched energy system, adding to its environmental footprint.

At the heart of AI's rapid expansion are data centers, vast facilities that power cloud services, machine-learning models and real-time data processing. In 2024, data centers accounted for about 1.5% of global electricity consumption, using approximately 415 terawatt-hours (TWh); a sharp increase from previous years and roughly equivalent to the entire electricity consumption of the UK. AI is accelerating this demand because its workloads are far more energy-intensive than traditional digital solutions. The need for specialized processors, round-the-clock cooling and continuous redundancy all contribute to a growing carbon footprint that frequently remains hidden. As local authorities expand their digital service delivery, the indirect carbon footprint of the tools they rely on grows too, further contributing to Scope 3 emissions.

Many tech companies and providers have made impressive public commitments to powering their data centers with 100% renewable energy, but the reality is far more complex. Renewable energy generation is intermittent, and large facilities often rely on fossil fuels during periods when clean energy is not available. Without transparent reporting, local authorities cannot fully understand the true climate impact of the digital services they procure.

Another environmental cost that often goes unnoticed is water consumption. Data centers use millions of liters of water every year for cooling. In 2022, global data centers used an estimated 1.105 billion liters of water per day. Based on current growth rates, 2024 consumption is likely over 1.249 billion liters of water per day, with projections exceeding 1.70 billion gallons of water/day by 2027.

GPT-3 alone may have consumed up to 700,000 L of clean water during training, according to research published in Nature Sustainability. Even everyday use contributes: a single 20–50 question conversation with ChatGPT could consume the equivalent of a 500 ml bottle of water, depending on cooling method and server location (UK Local Partnerships News 2025).

References

Australian Trade and Investment Commission (2025) Australia's data center landscape: Current and future potentials for AI and energy
Chosun Daily News (2025) Status in AI data centers
Downey B, McGuigan J (ed) (1999) Technocities. Sage Publications, London, pp 15–17
Noyman A (2025) Generative agents in the built environment
UK Local Partnership (2025) 9 September (personal service email)

Chapter 2
Emergence of AI New Urbanism as a Synthesis Science Where AI Meets Cities: AI Optimism for Data-Driven City Climate Planning

Kwi-Gon Kim

Abstract Cities around the world are being reshaped with AI optimism. We need to plan and operate smart and sustainable cities for smart climate solutions with AI. Simple AI urban planning and design initiatives can have extraordinarily powerful and positive consequences. New names for cities have arisen in conjunction with the usage of digital technology, namely, digital city, ubiquitous city, cyber cities, virtual cities, cloud cities, 5G-powered cities and predictable cities. AI urban philosophy provides the basics to develop an AI new urbanism approach. This chapter is oriented towards defining the AI urban philosophy and new urbanism approach. This is achieved by examining UN SDGs, criteria and indicators for sustainability, and suggesting indicator approaches, components and functions, as well as connecting data and AI platforms for prediction and optimization for decision making from a true synthesis science perspective. The chapter is organized according to these components. It involves suggestions for developing an 'AI-based urban sustainability optimizing model: Optimum sustainability with sustainable development criteria and indicators.'

Keywords Planning AI · Optimum sustainability · Sustainability cost and benefit analysis · AI urbanism · Climate-smart sustainability · Synergistic climate action · Future state of sustainability · AI apps

2.1 Emergence of a New Climate-Smart and Sustainable Eco-Urbanism: Urban Climate Agenda

Urban AI is an emerging concept that is an extension of the smart city paradigm, helping cities navigate the climate challenges of the twenty-first century.

K.-G. Kim (✉)
Seoul National University, Seoul, South Korea
e-mail: kwigon@snu.ac.kr

K.-G. Kim (ed.), *AI Transformation of Cities for Climate Solutions*,
https://doi.org/10.1007/978-3-032-21940-4_2

Some climate scientists believe the 1.5 °C limit of the Paris Agreement could even be breached before 2030. Cities are not only already experiencing the consequences of climate change, but also contributing to breaching the 1.5 °C target. Thus, city-planning policy should be connected with AI innovation to beat climate change.

Traditional concepts and definitions of urbanism are outmoded. The old urbanism was an urban design movement which promoted physically walkable neighborhoods containing a range of housing and job types without cyber connections and operation. It helped citizens reduce energy consumption and subsequently reduced greenhouse gas (GHG) emissions. It arose in the early 1980s and has gradually influenced many aspects of real-estate development, urban planning, and municipal land-use strategies without leveraging cyber-system-based solutions for carbon neutrality.

We are now concerned about adapting to climate change and the potential to mitigate climate change through AI technologies with well-planned AI cities.

However, there are problems here because the foundation of planning education and skills show the legacy of planning's origins in focusing largely on the physical mobility of people and goods within cities, and the location of physical facilities and land uses within a unitary, integrated city (Webber 1968).

Noyman et al. (2025) began by illustrating how traditional urban planning has often relied on top-down decision-making processes with limited insight into the actual experiences of urban dwellers. Historically, urban theorists like Jane Jacobs and Kevin Lynch emphasized the importance of understanding cities from the pedestrian's perspective. However, the tools available to planners were constrained by static models and pre-defined simulation frameworks. AI agents, Noyman argues, present a significant departure from these conventional approaches by enabling dynamic, real-time simulations of human movement, decision making and perception within urban environments.

As the AI industry continues to reshape the urban landscape, according to Noyman the challenge for planners, policymakers and researchers will be finding the right balance between automation and human judgment.

The insertion of smart climate sustainability into the city introduces today more dimensions into urban planning in terms of urban structure and function. In particular, the newly emerging urbanism tries to unlock the power of artificial intelligence with innovatory Generative AI. It is breakthrough news on synergistic climate action.

2.1.1 Potential Functions of AI Technologies for the Urban Climate Agenda

In the information age, businesses cannot succeed without effective information sharing between employees, departments and customers. This is true for governments as well; operating in isolation, governments cannot deliver efficient protections and value-added services to citizens. The effectiveness of government services depends directly on the quality of their interaction with citizens, business partners

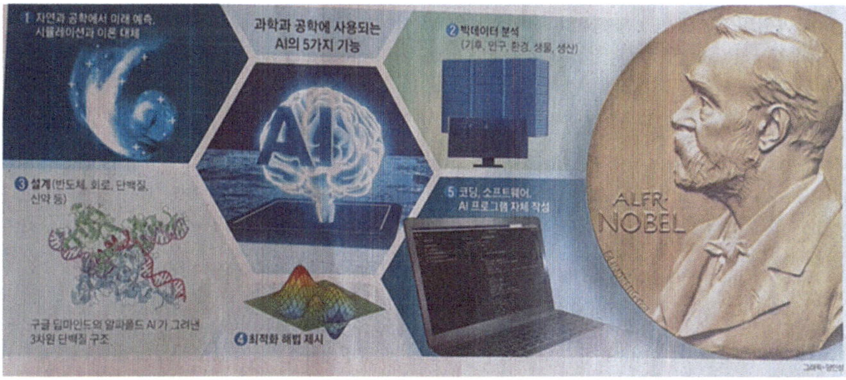

Fig. 2.1 Five functions of AI contributing to science and engineering. 2024 *Source* Chosun Daily News, December 18

and employees, in the context of agency and business processes. Implementing a smarter AI city must be considered as a process of integrating technology with the development process for starting, growing and transforming the city as time progresses. Initiatives should also look at improving the efficiency and efficacy of the city, remembering that humankind learns in the real world, but AI learns in a digital virtual world through mental and physical experiences (Kim 2025).

AI technologies have received wide attention in recent years as an innovation solution to various sectors such as: (1) future prediction and simulation; (2) big data analytics (climate, population, environment, biology, production); (3) design (semiconductors, circuits, proteins, new drugs, etc.); (4) optimization solutions and (5) coding, software and AI programming (Fig. 2.1).

AI has the potential to change human behavior and patterns beyond technical development. The evolvement of AI and the systems approach to planning have been very much influenced by physics. Planning AI could optimize urban planning and development processes and enhance efficiency with GenAI and physical AI.

AI servers (e.g. Google Cloud or Alibaba Cloud)) are finding their way into every aspect of city operations including public safety, planning, transportation and citizen service. This book examines the theory, methods and potential applications of AI in climate-responsive urban planning and decision making—exploring the combination of AI and city climate planning.

AI transformation (AIX) in the future has limitless applicability in cities. In the society of the future, AI is likely to be a key requirement for city competitiveness and citizens' quality of life. It is essential to create a city AI ecosystem for new goods and services. The establishment of AIX can accelerate the development of planning AI for the sustainability of cities. We need to uncover new ideas driving sustainability progress, and the more ambitious areas of adaptation to and mitigation of climate change.

Examples of application fields in city climate planning AI to optimize sustainability include the following themes:

- AI optimum urban sustainability model
- Urban land use (using land to mitigate climate change)
- Energy-saving urban form
- Land-based mitigation to reduce land-use emissions.

Scenarios that limit global warming to below 2°C by 2100 assume significant land-use change to support large-scale carbon dioxide (CO_2) removal from the atmosphere by afforestation/reforestation, avoidance of deforestation, and Biomass Energy with Carbon Capture and Storage (BECCS).

more sectors to be informed are as follows:

- Climate, environment
- Buildings
- Water
- Waste
- Energy
- Transportation
- Consumption/needs-based GHG emissions accounting
- Climate-responsive planning and design
- Civil engineering: Innovative bio-based construction strategies for near-zero energy buildings with low carbon footprints
- Blue-green infrastructure
- Carbon wetland system
- Human behavior
- Physical-social-cyber systems
- Innovative solutions for energy and water integration: From microgrids to city applications
- AI home
- NBS
- Ecosystem (habitat) health and integrity
- Grand design transformation
- Community participation: Citizen science-based media arts and visualization of data
- Climate impact assessment based on synergistic science

In addition to the potential examples above, AI technology should be extended to include more applications in cities.

As an example, the development and application of AI agents for optimization of stork habitats by integrating big data and AI platforms include the following steps:

(1) Collection and time series analysis of learning data for stork habitats:
(2) Literature review
(3) Review of case studies
(4) Field survey using AI sensors
(5) Use of stork AI servers (e.g. Google, Amazon, IBM, Naver, Daum)

(6) Determination of optimum stork habitat criteria in similar environmental circumstances
(7) Testing of deep learning programs to determine optimum habitats for storks in terms of mating, hiding, feeding, sleeping and daytime activities
(8) Application of the developed stork habitat AI agents and tools in the case study area
(9) Integration of human and machine intelligence for decision making

This approach would be a valuable addition to the definition of goals of AI habitat suitability evaluation as follows:

(1) Understand the fundamentals of AI and machine learning and their application in demand forecasting and optimization
(2) Implement techniques for preparing and preprocessing data for machine learning models
(3) Understand the role of time series analysis in demand forecasting
(4) Implement techniques for developing and evaluating regression and machine learning models for demand prediction
(5) Implement techniques for conducting scenario analysis
(6) Understand the legal and ethical considerations related to AI in habitat management
(7) Implement techniques for integrating AI models with existing habitat management systems
(8) Understand the economic and operational benefits of AI adoption in demand forecasting and optimization
(9) Develop strategies for managing AI-driven habitat management projects
(10) Develop strategies for measuring and analyzing the effectiveness of AI implementations.

More specifically, Table 2.1 shows the applicability of AI for urban policy, planning, program and project (4Ps) solutions for climate change. These include key points of the city agenda in the form of a synthesis towards AI creative disruption. The AI tools suggested in Table 2.1 can be combined with Table 1.1. so that AI can be integrated into different types of climate solutions (Table 1.1) for different levels of decision hierarchy (Table 2.1). For example, AI-based emission reductions can be occurred with government intervention such as mitigation policies (Table 2.1).

2.1.2 Types of New Cities: Autonomous Cities in the Urbanism of the 4th Industrial Revolution and 5th AI Business Age

As a consequence of digital transformation, new names for types of cities have emerged including the virtual community (Fig. 2.2), cyber city, cloud-based city (Fig. 2.3), 5G-powered city, U-city (ubiquitous city) and smart city.

Table 2.1 Urban 4Ps solutions for climate change: AI tools

4Ps	Descriptions	Applications of AI tools
Urban policy		Platforms like UrbanistAI, DeepBlocks, and TestFit for design and simulation GIS-based software like CityEngine for spatial analysis Platforms like UrbanSim[a] for strategic planning by simulating land use and transportation interactions Tools like FlyPix AI that use AI for satellite imagery analysis Generative models and AI agents that can model community behavior to predict the impact of policies
Urban planning		Platforms for generative design like TestFit and Spacemaker AI Platforms for sustainable neighborhood design such as Delve by Sidewalk Labs Tools that support community engagement like UrbanistAI AI-powered GIS for mapping and analysis, predictive modeling for traffic and resource allocation Tools for creating 3D visualizations and virtual simulations *Design and Analysis Tools* **TestFit** and **Spacemaker AI** (by Autodesk): These platforms focus on urban design and feasibility studies, generating schematic designs and floor plans for mixed-use developments *Delve by Sidewalk Labs*: This tool aids in designing sustainable neighborhoods by analyzing and predicting urban performance metrics *EvolveLAB's Veras*: An AI tool for architectural visualization and render enhancement, providing enhanced realism in design proposals *Archistar*: Used for site analysis and optimizing yields, helping planners understand the potential of a site *UrbanSim*[a]: This platform integrates land use, transportation, economy, and environmental factors to create models for simulating urban development Data Analysis and Prediction *GIS-based AI*: AI enhances Geographic Information Systems to analyze urban data, optimize infrastructure, and monitor environmental changes *Traffic and Mobility AI*: AI analyzes traffic patterns from sources like mobile location data to predict traffic flow and suggest infrastructure improvements *Environmental AI*: AI analyzes environmental data from satellite imagery and sensors to ensure the sustainability of urban projects Community Engagement and Visualization *UrbanistAI*: A platform designed to facilitate community engagement by allowing citizens to co-design urban spaces and visualize the impact of planning policies *Augmented Reality (AR) and Virtual Reality (VR)*: These technologies create virtual simulations and immersive experiences, allowing stakeholders to explore proposed developments in a realistic environment *Generative Design Platforms*: Tools like Hestia allow planners to prototype various building and site configurations, testing different parameters to explore design possibilities *Broader Applications* *Resource Allocation*: AI helps allocate resources like water, electricity and waste management more efficiently for sustainable city operations *Infrastructure Planning*: AI can predict the impact of new infrastructure projects and optimize the design of transportation systems

(continued)

Table 2.1 (continued)

4Ps	Descriptions	Applications of AI tools
Urban program		Not applicable
Urban project		Platforms for generative design, site analysis, sustainable design, visualization and community engagement, such as Spacemaker AI by Autodesk for site planning, TestFit for feasibility studies, Delve by Sidewalk Labs for neighborhood design, Hypar for parametric workflows, and Adobe Firefly for image generation Other key tools are Digital Blue Foam (DBF) for generative design, Aino for advanced site analysis, Archistar for site analysis and yield optimization, and UrbanistAI for public engagement

Source Google Gemini (responses), 2025 (modified by the author)

[a]UrbanSim Inc. helps urban development stakeholders make smarter plans with less effort. State-of-the-art urban simulation, 3D visualization, and shared open data empowers users to explore, gain insights into, and develop and evaluate alternative plans to improve their communities. The UrbanSim's platform incorporates interactions between land use, transportation, the economy, and the environment to build better cities.

The AI city is an innovative city model which optimizes and improves diverse urban systems and services by using AI agents and technologies. The AI city is far beyond the concept of smart cities. It pursues the creation of more intelligent, efficient and sustainable urban environments by applying AI agents and tools in every aspect of the city.

In particular it seeks to solve equations necessary for design, and performance prediction and optimization processes (Fig. 2.4).

In this section, the main features of AI cities include:

- Urban data- and AI agents-based city operation and optimization focusing on environment and climate
- Data and AI agents-based decision making and prediction of future city image from synthesis science perspectives
- Integration of SDGs with city climate planning
- Promotion of citizen, science-based, local participation in the entire process of urban planning and development
- Combined use of deep learning work flow together with urban and ecological modeling processes
- Provision of hardware infrastructure (e.g. AI server, AI chip, GPU, etc.).

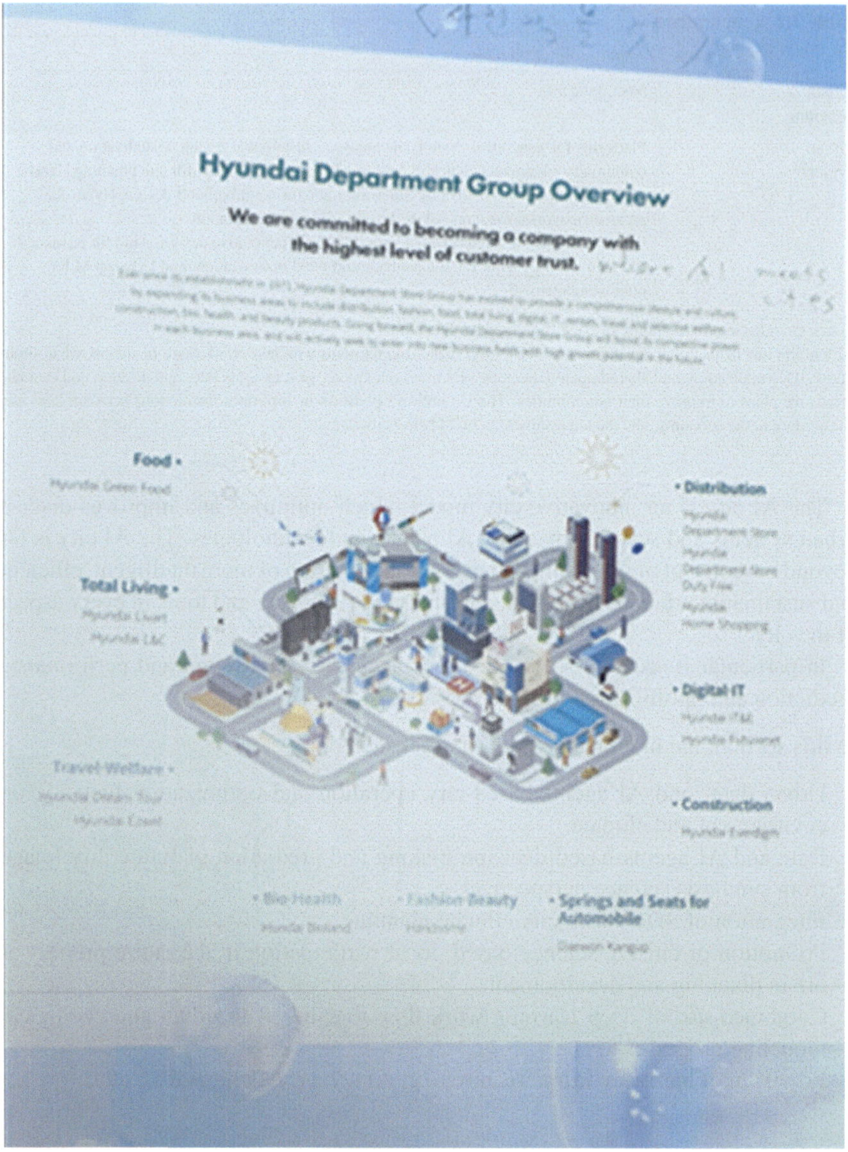

Fig. 2.2 AI virtual city model

Fig. 2.3 Cloud-based city

Fig. 2.4 AI research tool

2.2 UN SDGs, Criteria and Indicators for Climate Action Plans, Sustainability Appraisal and Climate Impact Assessment: Connecting Local and National Climate Action

Sustainable Development Goals (SDGs) should be at the top of the AI agenda to optimize sustainability in cities. This section aims to address the pressing challenges and AI opportunities that urban areas face in the context of sustainability. We need to change our way of thinking about AI technologies. This section is part of the AI SDG research on optimum or desirable sustainability discussed in this book.

Many sustainability indicator research studies have been focused on the assessment of the state and trends in the applications of indicators for evaluating the current level of sustainability on a comparative basis (Hak et al. 2007). This is done through comparison of relative performance evaluation, rather than attempting to explore future planning indicators. The research on future planning indicators for urban sustainability can be considered as a contribution to extend the body of knowledge in an innovative AI agent research stream.

Backcasting is used for many city climate action plans to envision a city with zero carbon emissions by 2050. The backcasting process identifies the necessary shifts in energy infrastructure, transportation systems, building regulations and lifestyle changes required to reach that goal. The AI demand/consumption planning approach is more realistic and dynamic than the traditional backcasting approach to climate action plans based on sustainability indicators.

AI demand planning is the use of artificial intelligence, particularly machine learning, to forecast future product or service demand by analyzing large volumes of real-time and historical data, external factors and market trends. This method is more accurate and dynamic than traditional forecasting, leading to benefits like improved CO_2 emission inventory management, and increased responsiveness to climate change.

It works as follows:

(1) Data analysis

AI systems analyze diverse datasets, including historical sales, market signals, weather patterns, social media trends and environmental, social and economic indicators to identify complex patterns and correlations.

(2) Machine learning algorithms

Algorithms like time series analysis, regression modeling and neural networks are trained on this data to create predictive models that learn and adapt over time.

(3) Dynamic forecasting

Unlike static traditional methods, AI forecasts are dynamic and can be updated in real time as new data becomes available, allowing for immediate adjustments to changing circumstances.

Box 2.1

The backcasting approach is a strategic planning method that begins by defining a desirable future state and then works backward to identify the specific actions, policies, and changes needed to achieve it. Unlike forecasting, which projects current trends forward, backcasting focuses on creating a visionary future and reverse-engineering the steps to get there. This method is particularly useful for tackling complex, long-term challenges, such as sustainability goals, and promotes systems thinking and collaboration among stakeholders to achieve the vision.

How the backcasting approach works:

(1) **Define a desirable future**:

Start by envisioning an ideal, often long-term, future state. This vision serves as the target endpoint.

(2) **Identify necessary steps**:

Work backward from that future goal, identifying the key changes in law, culture, and technology that would need to occur to make the future a reality.

(3) **Determine present actions**:

Identify the current actions, policies, programs, and resources that are necessary to connect the future vision to the present situation.

(4) **Assess feasibility**:

Evaluate the steps and potential obstacles, considering unintended consequences and trade-offs.

Figure 2.5 shows the relationship between NDCs, LDCs and EDCs for the implementation of SDGs. The definitions of NDCs, LDCs and EDCs are also included in the same figure. This figure explains how to downscale in all climate actions.

The Sustainable Development Goals (SDGs) are a universal call to action adopted by the United Nations in 2015 to end poverty, protect the planet and ensure peace and prosperity for all. There are 17 goals with 169 targets and 230 unique indicators, covering a wide range of environmental, social and economic development issues.

The 17 SDGs are a framework for global action, addressing issues like poverty, hunger, health, education, climate change and social justice. Each goal has specific targets, which are measurable actions aimed at achieving the broader goal. The SDGs are designed to be measurable, and progress is tracked using a specific set of indicators. These are specific metrics used to monitor progress towards each target and goal.

Goal 13 is concerned with climate change: **Take urgent action to combat climate change and its impacts.**

The targets and indicators for Goal 13 are as follows:

13.1. Strengthen resilience and adaptive capacity to climate-related hazards and natural disasters in all countries

  13.1.1. Number of countries with national and local disaster risk reduction strategies

  13.1.2. Number of deaths, missing persons and persons affected by disaster per 100,000 people

13.2. Integrate climate change measures into national policies, strategies and planning

  13.2.1. Number of countries that have communicated the establishment or operationalization of an integrated policy/strategy/plan which increases their ability to adapt to the adverse impacts of climate

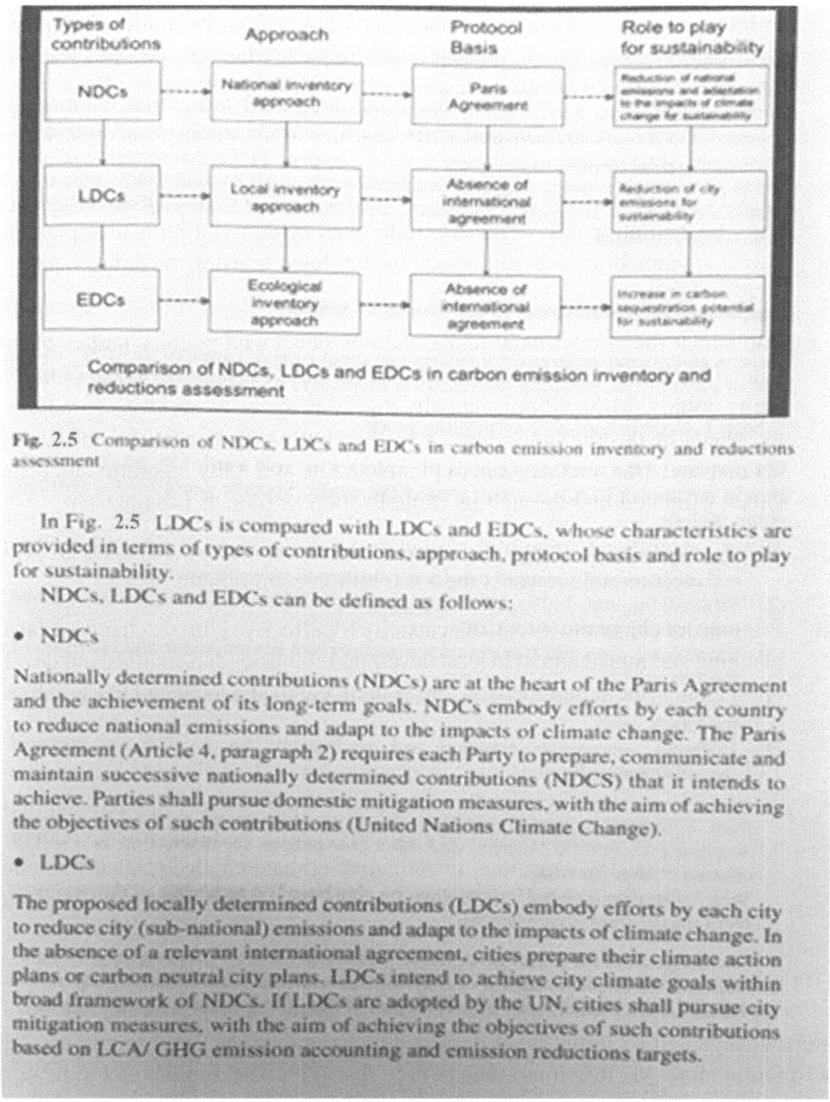

Fig. 2.5 Comparison of NDCs, LDCs and EDCs in carbon emission inventory and reductions assessment

In Fig. 2.5 LDCs is compared with LDCs and EDCs, whose characteristics are provided in terms of types of contributions, approach, protocol basis and role to play for sustainability.

NDCs, LDCs and EDCs can be defined as follows:

- NDCs

Nationally determined contributions (NDCs) are at the heart of the Paris Agreement and the achievement of its long-term goals. NDCs embody efforts by each country to reduce national emissions and adapt to the impacts of climate change. The Paris Agreement (Article 4, paragraph 2) requires each Party to prepare, communicate and maintain successive nationally determined contributions (NDCS) that it intends to achieve. Parties shall pursue domestic mitigation measures, with the aim of achieving the objectives of such contributions (United Nations Climate Change).

- LDCs

The proposed locally determined contributions (LDCs) embody efforts by each city to reduce city (sub-national) emissions and adapt to the impacts of climate change. In the absence of a relevant international agreement, cities prepare their climate action plans or carbon neutral city plans. LDCs intend to achieve city climate goals within broad framework of NDCs. If LDCs are adopted by the UN, cities shall pursue city mitigation measures, with the aim of achieving the objectives of such contributions based on LCA/ GHG emission accounting and emission reductions targets.

Fig. 2.5 Relationship between NDCs, LDCs and EDCs for the implementation of SDGs. *Source* Kim (2025) p.355

change, and foster climate resilience and low greenhouse gas emissions development in a manner that does not threaten food production (including a national adaptation plan, nationally determined contribution, national communication, biennial update report or other)

13.3. Improve education, awareness-raising and human and institutional capacity on climate change mitigation, adaptation, impact reduction and early warning

13.3.1. Number of countries that have integrated mitigation, adaptation, impact reduction and early warning into primary, secondary and tertiary curricula

13.3.2. Number of countries that have communicated the strengthening of institutional, systemic and individual capacity-building to implement adaptation, mitigation and technology transfer, and development actions

13.a. Implement the commitment undertaken by developed-country parties to the United Nations Framework Convention on Climate Change to a goal of mobilizing jointly $100 billion annually by 2020 from all sources to address the needs of developing countries in the context of meaningful mitigation actions and transparency on implementation and fully operationalize the Green Climate Fund through its capitalization as soon as possible.

13.a.1. Mobilized amount of United States dollars per year starting in 2020 accountable towards the $100 billion commitment

13.b. Promote mechanisms for raising capacity for effective climate change-related planning and management in least developed countries and small island developing States, including focusing on women, youth and local and marginalized communities.

13.b.1. Number of least developed countries and small island developing States that are receiving specialized support, and amount of support, including finance, technology and capacity-building, for mechanisms for raising capacities for effective climate change-related planning and management, including focusing on women, youth and local and marginalized communities.

An example of a simple checklist for climate change with information provided by UN SDGs is shown in Table 2.2. Metrics of goals, targets, and indicators for climate change and its impacts are designed to encourage the rapid increase in the use of sustainable AI. It is important to note that the goals, targets and indicators have to be downscaled so that these are applicable to cities. Furthermore, climate standards could add more value to AI.

These sustainability indicators are used for AI modeling of climate impact assessment, AI sustainability cost–benefit analysis, goal achievement matrix for plan evaluation and multi-AI agents in this book.

Table 2.2 Metrics of goals, targets and indicators for climate change and its impacts for sustainability cost and benefit of climate action

Goals	Targets	Indicators	Values
Goal 13	1. Resilience and adaptive capacity	1.1. Number of local disaster risk reduction strategies	
	2. Integration of climate change measures into policies, strategies and planning	2.1. Establishment of an integrated climate policy/strategy/plan	
	3. Improvement of education, awareness-raising and human–institutional capacity on climate change mitigation, adaptation, impact reduction and early warning	3.1. Establishment of mitigation, adaptation, impact reduction and early warning into primary, secondary and tertiary curricula	

Note Sustainability indicators as a contribution to extend the body of knowledge in an innovative AI agent research stream

2.3 Urban Sustainability Transformation

AI technologies should be utilized for urban sustainability transformation to solve global problems such as climate crisis.

Sustainability transformation refers to a fundamental, system-wide shift across technological, economic and social factors to ensure long-term human well-being and sustainable development. It involves integrating sustainable practices into a city's policy, planning, programs and projects to create positive environmental, social, and economic impacts. This transformation goes beyond incremental changes and requires a holistic, long-term perspective that considers the impact of a city's activities on stakeholders and society.

Sustainability transformation addresses environmental, social, and governance (ESG) challenges, such as climate change, resource depletion, social inequality and ethical governance. The urgency of addressing climate change is a major driver of sustainability transformation.

The concept of AI agents and sustainability transformation should be integrated to achieve twin transformation. This book tries to illustrate the integration using climate change initiatives. Key aspects of integrating AI agents and sustainability transformation from climate change perspectives are as follows:

- LCA transformation: AI technologies should be applied in the entire process of sustainability transformation. Major actions significantly affecting the quality of the sustainability transformation should be included to provide decision makers more complete information and to enable better realization of the transformation.
- Adaptive transformation: AI should play a role of an aid for sustainability transformation of cities adapting to new learning data and technology.
- Indicator transformation: Urban sustainability indicators should be developed and applied in planning AI.

Box 2.2
Use of Indicators for Mitigation and Adaptation Policies

As an example, indicators of carbon storage in forests and other land uses have gained much prominence in evaluating and endorsing land-based climate change mitigation policies. The outcomes of such assessments can have direct livelihood implications for dwellers living at the forest–agriculture frontier, such as shifting cultivators.

Here, we suggest ways of connecting the UN's Agenda for Development: "The dimensions of development" with AI solutions:

- Peace as the Foundation
- The Economy as the Engine of Progress
- The Environment as a Basis for Sustainability
- Justice as a Pillar of Society
- Democracy as a Good Governance

Article 6 of the Paris Climate Agreement recognizes that some parties choose to pursue voluntary cooperation in the implementation of their nationally determined contributions (NDCs) to allow for higher ambition in their mitigation and adaptation actions and to promote sustainable development and environmental integrity.

AI can be used to create agents which connect the implementation of Article 6 of the Paris Agreement with sustainability transformation in an integrated holistic manner.

2.4 Sustainability Indicator Approach to Prediction and Optimization from a Synthesis Science Perspective

This section is addressed to data needs and associated technology for predicting and optimizing impacts of proposed actions on the climate with AI agents.

Synthesis science provides a synthesis view of the existing conditions and trends that provides information for decision making in the relevant sector. It helps multi-indicator approach to the so-called multi-AI agents for prediction and optimization for decision making.

Box 2.3
Sustainability Synthesis Working Groups

Joint International Synthesis Call on Interactions between Biodiversity and Climate Change

As part of a joint call from a consortium of international synthesis centers around the theme of interactions between biodiversity and climate change, the **Just Transformations for Sustainability Initiative (JTS) at University of Notre Dame** and the **Institute for Global Change Biology (IGCB) at the University of Michigan**, announce a new open call for proposals to form **Synthesis Working Groups**.

The JTS Initiative and IGCB joint call will support two types of Synthesis Working Groups: 1) Joint IGCB–JTS Initiative Working Groups and 2) Joint IGCB–JTS Working Groups with other Synthesis Centers. This call for proposals is for working groups that will commence in **late 2025 or early 2026.**

The sustainability indicator approach is a process of considering the effects of a current or proposed plan (with more than one option) on a range of sustainability indicators (or criteria), in order to improve its performance or identify the most sustainable option. The optimization of planning activities requires the use of objectives or indicators in models in planning. Planning objectives state what is intended. The plan's performance against objectives is normally measured by using indicators. The environmental or climate effects of the plan or project can be described, analyzed, compared and optimized with indicators by AI means.

AI can play an important role in encouraging a more sustainable approach to climate change mitigation and adaptation through planning with sustainability objectives and indicators under the global SDGs. UN SDGs suggest criteria and indicators of sustainable development, as described in the previous sections. These indicators can be used for the integrated consideration of environmental, social and economic sustainability for urban planning and decision making within the broad concept of SDGs. Analyzing sustainability indicators is essential for optimal planning and reducing GHG emissions. A major feature of this approach is that optimizing indicators are expressed as a future state of sustainability.

AI sustainability studies should be carried out to explain the future of sustainability rather than describing the current status of sustainability by using sustainability indicators (Newman and Kenworthy 1999).

The objectives and criteria for optimizing climate action are set out at different levels, as shown in Table 2.2. Even though urban planning and development processes are technically very complex, two significant components of progress are prediction and optimization, which can be improved by AI with tangible impacts. The use of AI can aid planners and developers in modeling future change based on an indicator approach and optimizing climate-smart and sustainable eco-city solutions within the bounds of data availability.

The sustainability indicator system can also be developed in carbon inventory and monitoring for optimizing climate action. Carbon inventory and monitoring is critical for the reporting and verification of carbon stocks and change. AI-powered tools improve individuals' and organizations' carbon footprint tracking and analysis. With carbon-emission reporting, these innovations provide unprecedented insights into carbon impacts, enabling more effective carbon strategies. How does AI achieve this?

- Personal carbon tracking: AI apps and edge AI (local AI: phones, etc.) help to analyze user behavior and purchases, and consumption of goods and services to estimate individual carbon footprints and suggest personalized reduction strategies. Personalized AI can change a GHG accounting system from production-based accounting to needs/consumption-based accounting.
- Corporate emissions monitoring: AI systems integrate data from various sources to give businesses accurate, real-time carbon emission calculations.
- Supply chain analysis: AI models assess the carbon impact of entire supply chains, identifying hotspots for emissions reduction and optimizing logistics.

Examples of AI online computer modeling programs with indicators for different sectors and subsectors that enable planners to enter the AI market include:

- Mathematical modeling of the real world (AI computer algorithms for problem-solving and decision making)
- Visual presentation of the real world: Visualization of data
- Digital twin
- Metaverse
- Space intelligent technology
- Digital shadow methods.

These examples will be further elaborated in the following chapters.

References

Hak T, Moldan B, Dahl AL (eds) (2007) Sustainability Indicators. Island Press, Washington, pp 56–57

Kim KG (2025) LCA and consumption/needs-based GHG accounting for climate action: a pathway to carbon neutrality. Springer, Heidelberg, p 355

Kim JH (2024) Five functions of AI used for science and engineering. Chosun Daily News, 18 December

Newman P, Kenworthy J (1999) Sustainability and cities: overcoming automobile dependence. Island Press, Washington, p 11

Noyman A et al. (2025) TravelAgent: Generative agents in the built environment. Environ Plann B Urban Anal City Sci, https://doi.org/10.1177/23998083251136048

Webber M (1968) The post-city age, Daedalus. Autumn, pp 1381–1421

Part II
Theoretical Frameworks and Methodologies for AI Transformation of Cities, Strategies and Action Plans

Chapter 3
Framework for Combining AI with the Life Cycle Process of Urban Planning and Development Process as a Whole: A Paradigm Shift

Kwi-Gon Kim

Abstract Cities are responsible for a significant share of global emissions, but they also offer the highest leverage for climate mitigation and adaptation. Urban planning directly shapes how people move, how buildings use energy and how resilient communities respond to heat and flooding. A climate-aligned urban plan integrates five layers: (1) urban form (shape of the city), (2) infrastructure systems, (3) building, (4) land-use policy and (5) community and environmental resilience. This integrated approach simultaneously reduces emissions, improves quality of life and strengthens resilience to climate impacts. However, lack of original innovation in theories and frameworks often follows trends. The theoretical application of AI in cities require an all-inclusive and cooperative framework. All-inclusive denotes that AI activity and tools can be incorporated into the whole urban planning and development process, including the entire life cycle of the planning, construction, operation, monitoring and evaluation by using AI-generated future city images. Inputs are required from many disciplines to ensure that a comprehensive analysis of the applications has been accomplished. This chapter describes a rational framework for usage of AI in the urban planning and development process. The process includes a planning and design phase, (data collection, data storage, data analytics, data communication (5G/6G), data service, expert prediction, SWOT analysis, master plan and specific sectoral plan preparation, detailed plan and working drawings and implementation action plan), a construction phase, an operations and maintenance phase and a monitoring and evaluation phase. The chapter develops a matrix format for describing the relationship between the life cycle phase of planning and development activities and AI technologies.

Keywords AI platform · AI-driven city · Sustainability cost–benefit analysis · Data sensing · Data analytics · Constraint · AI model · Carbon footprint reduction · Edge AI · Climate impact assessment · Habitat system · AI finance · Kitakyushu

K.-G. Kim (✉)
Seoul National University, Seoul, South Korea
e-mail: kwigon@snu.ac.kr

K.-G. Kim (ed.), *AI Transformation of Cities for Climate Solutions*,
https://doi.org/10.1007/978-3-032-21940-4_3

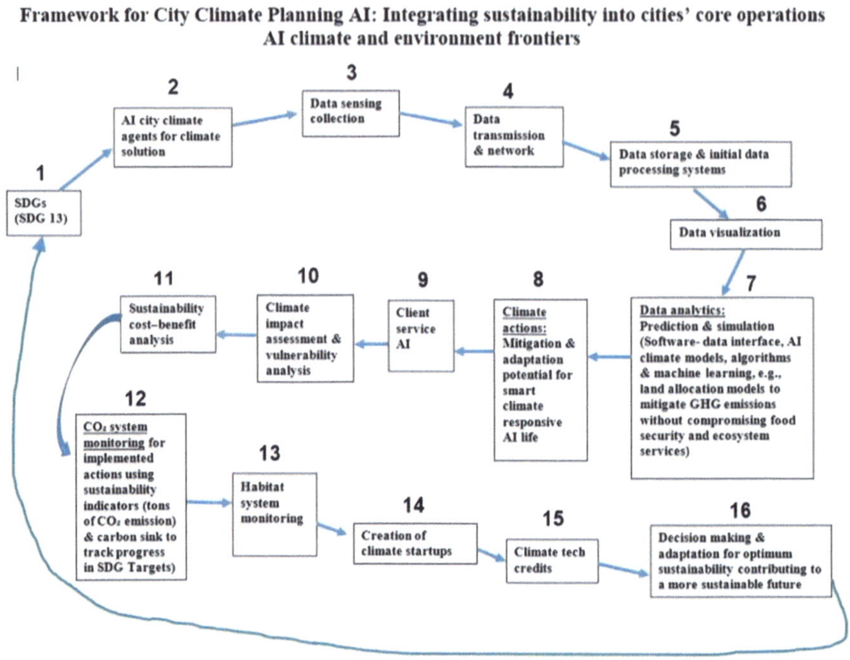

Fig. 3.1 Framework for AI city climate planning and assessment

3.1 Framework for Potential Usage of AI Technologies in the Climate-Responsive Urban Planning and Development Process

There is a growing acceptance that a downscaled climate framework can help city climate planners to understand and, in appropriate circumstances, predict and optimize the behavior of urban systems. However, it is important for the potential user to appreciate the place of the components of the framework within the wider context of the urban planning and development process (Lee 1974).

Figure 3.1 shows a rational framework for conducting a city climate planning and AI integration, from sustainability cost–benefit perspectives, for a positive footprint.

3.2 Steps in the Framework

Core steps of the framework shown in Fig. 3.1 are described below. The major step in the AI city climate planning process that requires the greatest degree of scientific application of AI technology is the prediction and optimization of climate solutions.

3.2.1 Step 1: Data Sensing and Collection

Data sensing and collection is the process of using physical devices, such as sensors, to detect, measure, and gather information from the real world and then transmitting it for analysis, storage, and use in decision-making. In the Internet of Things (IoT) context, sensors collect parameters like CO_2 emissions, temperature, humidity, pressure, animal movement, human behavior or motion, which are then often converted to digital data, sent to processing platforms via Wireless Sensor Networks (WSNs), and utilized for purposes like process optimization, condition monitoring, and enabling predictive maintenance.

3.2.2 Step 2: Data Analytics–Prediction and Optimization

Data analytics includes the land-allocation model, the smart energy-grid model, the smart wetlands-grid model (or connectivity-scoring model) and the climate model. The purpose of the data analytics step is to make predictions and optimization, and drive informed decision making.

Another purpose of the data analytics model is the evaluation of climate change mitigation and adaptation measures by using AI creative disruption technologies. Figure 3.2 shows a conceptual framework for a land-use system as an AI agent model for assessing and optimizing the potential of land-based measures for eco-cultural tourism.

3.2.3 Step 3: Climate Actions–Mitigation and Adaptation Potential for Smart Climate-Responsive AI Life Cycle

This step explores the nexus between the AI life cycle and the pursuit of net-zero emissions, highlighting the potential of AI in driving sustainable development and combating climate change. The step examines various climate actions which include AI applications for mitigation and adaptation measures. This review underscores the transformative potential of AI in achieving a sustainable, net-zero future and provides insights for policymakers, researchers and practitioners engaged inclimate-change mitigation and adaptation.

A summary of the combined technologies of climate-change measures and AI for smarter and greener efficiency-driven solutions, including AI technologies, carbon footprint reduction and resilience, is presented in Table 3.1.

AI enables to achieve both mitigation and adaptation. The World Resources Institute (WRI 2025) recently analyzed more than 300 adaptation investments and found that over half of them are also expected to reduce GHG emissions in the long run; for example, by adopting climate-smart farming practices, bringing trees and green spaces to cities, or introducing more resilient renewable energy sources.

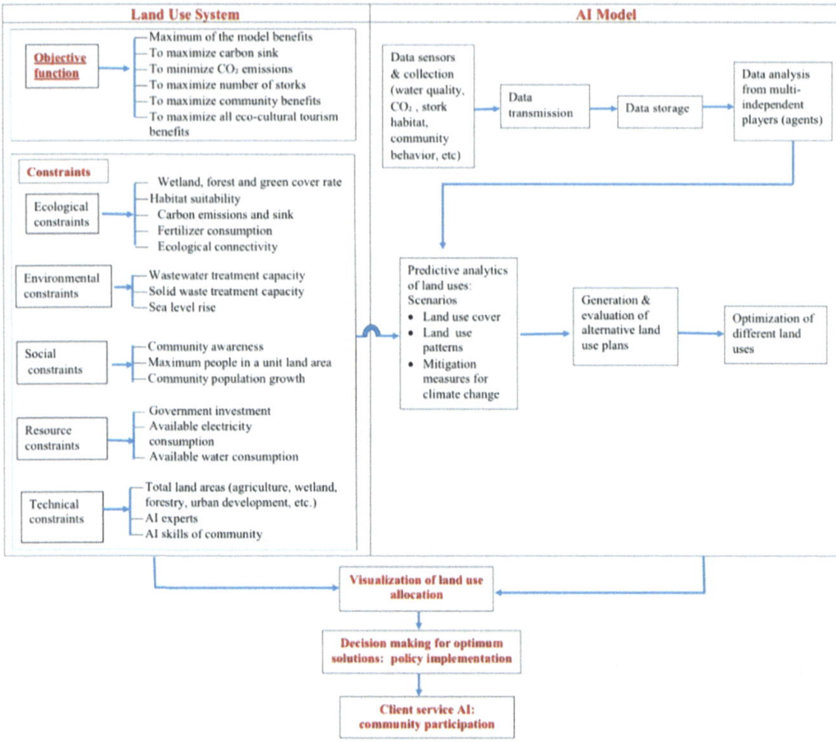

Fig. 3.2 AI agent model for assessing and optimizing the potential of land-based measures. *Notes* 1. Technical constraints also include AI model licensing, customization and post-construction operation and maintenance. 2. The available AI tools include Landstack AI and UrbanSIM

Strategies like growing trees can help curb climate change while at the same time helping to buffer communities against its impacts, such as drought and flooding.

Below are six examples of actions that can curb emissions in high-polluting sectors of the global economy while also improving our ability to deal with the worsening climate impacts:

(1) Expand decentralized energy
(2) Promote sustainable land Use
(3) Invest in climate-smart buildings
(4) Improve mass transit
(5) Protect and restore terrestrial and coastal wetlands
(6) Secure indigenous peoples' land rights.

These illustrate climate actions that address both adaptation and mitigation simultaneously. Many other examples exist and can accommodate different sectors, ecosystems, countries and communities. The climate crisis is massive and urgent and, given limited funding and capacity, policymakers need to prioritize integrated efforts like these to improve people's lives and protect the planet by AI means.

Table 3.1 Summary of combined technologies of climate-change measures and AI for smarter and greener efficiency-driven solutions

Categories of climate-change measures	Methods and techniques	AI technology applications: AI frontiers
Mitigation	Simulation of population growth • Urban planning – Transit-oriented development • Infill-densification • Urban blue-green infrastructure • Land use – Long-term forecasting – Land use pattern • Long-term consequences of zoning decisions • Urban form – Compact city – 15-min city • Energy demand control technology • Integrated renewable energy • Electricity grid systems Buildings • Integrated design process (IDP) into urban planning • Optimizing energy efficiency: • District heating/cooling Waste • Improving waste management • Alternate waste treatment • Waste to energy Water • Water quality maintenance • Saving water consumption Transportation • Enabling next generation mobility • Low carbon mobility options • Travel demand management • Electric cars with autopilots Lifestyle • Behavior and lifestyle changes • Personal carbon allowance • Municipal clean energy procurement	AI-powered disruption AI agents for data processing and model execution for forecasting population growth Predictive analytics (time series analysis) AI water and green space grid system AI innovation to ensure that its development aligns with sustainable goals AI sustainable goal achievement matrix model AI land allocation model AI EIA and climate impact assessment AI and machine learning (deep learning) for predictive energy demand forecasting and optimization algorithms Optimization of smart grid projects and energy use Optimization of renewable energy production by monitoring electric equipment with connected sensors Employment of sensors and machine learning to monitor and adapt energy consumption in buildings and industries, thereby improving energy efficiency for efficiency-driven solutions Optimization of waste treatment and maximizing material and energy recovery from waste AI tele-monitering station (TMS) Car energy demand prediction Traffic jam management AI traveller's guide AI autopilots Personalized AI

(continued)

Table 3.1 (continued)

Categories of climate-change measures	Methods and techniques	AI technology applications: AI frontiers
Adaptation	Flooding Earthquake Landslides Drought	AI-driven warning system AI in earthquake prediction AI rock crack gauger Infrastructure stress points: defect and risk assessment, and optimization
Integration: cross-cutting climate action measures	Climate change as a city policy Climate change planning and initiatives promotion Clean Development Mechanism (CDM) Database Survey and research Awareness raising and information provision Education and training Recognition improvement Demand and consumption of goods and services Supply chain management Public safety Edge AI (local AI) AI campus	Data infrastructure of AI for climate change: Big data and AI data center Analysis of data for environmental monitoring such as mapping CO_2 and methane emissions, and biodiversity Modern data center and cloud initiatives AI-powered Chatbot and ChatGPT, and virtual assistants to handle citizen inquiries and provide personal information AI human behavioral analytics and visual media art Visual simulation Time series analysis in demand forecasting Regression and machine learning models for demand prediction Optimization algorithms and their application in supply chain management AI CCTV camera AI systems that perform data processing and model execution directly on local devices—such as smart phones, laptops or servers without cloud-based infrastructure AI platform for learning and developing AI skills, emphasizing open access and free education

3.2.4 Step 4: Climate Impact Assessment and Risk Vulnerability Analysis

A climate impact assessment (CIA) identifies and quantifies the expected effects of climate change on a system, while a climate vulnerability analysis assesses a system's sensitivity and adaptability to these impacts to understand its overall risk. These processes, often combined as Climate Risk and Vulnerability Assessments (CRVAs), help decision makers, communities and organizations understand future climate risks, identify areas for action and develop effective adaptation strategies to build resilience against climate change.

At the project level, CRVAs can be used for assessing the impacts of proposed projects on climate change. Table 3.2 focuses mainly on the assessment process at the project level. AI climate impact assessment could become part of the Greenhouse Gas Protocol to advance AI-driven climate solutions in cities.

3.2.5 Step 5: Sustainability Cost-Benefit Analysis for Mitigation, Adaptation and Cross-Cutting Themes

A sustainability cost–benefit analysis (SCBA) (or Sustainable Development Cost–Benefit Analysis) extends traditional cost–benefit analysis (CBA) to evaluate projects by including economic, environmental and social costs and benefits in a comprehensive analysis from sustainability perspectives. The process involves identifying these diverse costs and benefits over a long-term timeframe, assigning monetary values where possible (though challenging for climate, environmental and social factors), discounting future values to their present value and weighing the total present value of benefits against costs to guide decision making towards more sustainable and efficient options.

Key steps in a sustainability cost–benefit analysis include:

(1) Identify costs and benefits
(2) Assign monetary values
(3) Discount future values
(4) Calculate net present value
(5) Assess risk and uncertainty
(6) Make a decision.

If the present value of benefits sufficiently exceeds the present value of costs, it strengthens the case for the sustainability initiative.

Table 3.3 shows a sustainability cost–benefit analysis table template that can be used to evaluate the financial and non-financial impacts of proposed climate actions.

Table 3.2 AI applications in climate impact assessment at the project level

CIA stages	Descriptions	AI applications (AI software programs)	Attainable or unattainable
1. Description of the project proposal	Location Size Time period Method of construction		O Knowledge gaps identified in the AI transformation of CIA
2. Description of the existing environment	Types of data: Physical–chemical data Biological data Cultural data Socioeconomic data Data analysis Remote sensing Computer-mapping approaches Data storage and retrieval systems	AI data center AI data analysis AI data platform Data security IBM Watson (data insights) Tableau (data visualization) Microsoft Azure storage Virtual Machine (VM) Cloud services and HDInsight Ceph storage (Ceph Multi-cloud K8s) Amazon Web Services	The deployment of CO_2 sensors may be too sparse to report the existing CO_2 situation in the site
3. Impact prediction	The primary and secondary effects of project development and implementation Identification of environmental modifications that may be significant Forecasting of the quantity and/or spatial dimensions of changes in environment identified Estimation of the probability that the impact (environmental and climate change) will occur (time period)	Urban simulation 3D visualization AI land use change models: multiple regression analysis 1.5 °C climate-change modeling SDA solution NVIDIA Jetson Altera FPGA system NUBISON AIOT IBM Watson (predictive analytics) Google Cloud's VertexAI (machine learning models) Microsoft Copilot Studio (customization of Copilot for specific needs)	Without climate big data, deep learning-based predictive models may not be useful Other sensing technologies such as crowdsourcing-based applications using mobile phones could be used

(continued)

Table 3.2 (continued)

CIA stages	Descriptions	AI applications (AI software programs)	Attainable or unattainable
4. Assessment (evaluation)	Extended cost–benefit analysis to quantify the intangible and long-term costs and benefits of a particular project and estimate meaningfully discount rates and opportunity costs Land-carrying capacity analysis using climate sustainability indicators Extension of risk analysis Assessing the cumulative impacts of new land uses Quantitative and qualitative standard assessment	Microsoft Copilot Studio (customization of Copilot for specific needs)	
5. Selection of mitigation measures and mitigation-based alternatives: optimization	Incorporating environmental and climate factors into planning process Checklist methods Matrices methods Network analysis methods Project alternative evaluation methods Sustainability cost–benefit analysis Goal achievement matrix	Autodesk InfraWorks AI impact operation management AI goal achievement matrix using sustainability indicators	It is of great importance to the sustainability of the project if we can optimize mitigation measures over the entire life cycle of the project, including supply chain emission through smarter procurement towards net zero
6. Climate monitoring	Climate data collection Climate data processing Climate data analysis	Integrated AI system for performance verification for impact assessment Envirosuite IBM Watson Environmental Compliance Google Earth Engine EarthAI Autodesk InfraWorks	

(continued)

Table 3.2 (continued)

CIA stages	Descriptions	AI applications (AI software programs)	Attainable or unattainable
7. Public participation	Incorporating of public participation through the process leading to a decision regarding a proposed action Forward and feedback techniques	Perplexity (citizen science) Client service AI Personalized service AI Use of chatbot and AI-aided visual methods for presentations for residents on climate impacts	

Table 3.3 AI sustainability cost–benefit analysis model (format)

Category	Climate action initiatives	Details (Project)	Initial cost (Year 0) plus ongoing cost (Year 1–Year 30) ($/year)	Short-term benefits (Year 0–Year 15)	Long-term benefits (Year 16–Year 30)	Net Present Value (NPV) and Benefit/Cost Ratio (BCR)
Mitigation	A (renewable energy) B C	A-1 (solar panel installation) A-2 (wind power generation) B-1 B-2				
Adaptation	A B C	A-1 A-2 B-1 B-2				
Cross-cutting measures	A B C	A-1 A-2 B-1 B-2				
Total (overall sum)						

3.2.6 Step 6: AI CO_2 System Monitoring

AI CO_2 system monitoring involves using sensors to detect carbon dioxide levels, which is crucial for health, safety and environmental applications. In industrial settings, monitoring prevents dangerous gas exposure in areas with high CO_2 concentrations. For indoor environments like offices and schools, it ensures good ventilation, improves cognitive function and productivity and can help reduce virus transmission by indicating poor air quality. In agricultural and environmental research, monitoring supports plant growth, crop preservation and climate change studies.

CO_2 monitoring works as follows:

- Sensors and detectors: Devices like CO2 Meter or Aerosphere sensors are placed in an area to measure carbon dioxide concentrations.
- Parts per million (ppm): Monitors measure CO_2 levels in parts per million (ppm) to provide a quantifiable reading of air quality.
- Alarms and displays: Many systems have two parts: a sensor placed near potential CO_2 sources (like a tank) and a remote display mounted at eye level for easy reading and early warning.
- Ventilation and control: When CO_2 levels rise, indicating poor ventilation, the system can trigger an alert or activate a ventilation system to bring in fresh air.

Applications of CO_2 system monitoring include:

- Environmental monitoring: Systems monitor greenhouse gas levels in the atmosphere to track climate change and support policy decisions.
- **Carbon capture, utilization, and** storage (CCUS): Monitoring is critical for managing the capture, transport, and storage of CO_2 to protect infrastructure and ensure safe permanent storage underground.
- Climate change mitigation: Provides data to monitor, reduce greenhouse gas emissions and prepare LCA/GHG accounting, contributing to global climate efforts.
- Obtain climate tech credits for climate change.

3.2.7 Step 7: Habitat System Monitoring

Habitat system monitoring is needed for the habitat suitability analysis described in Fig. 3.1 (Treweek 1999).

Habitat system monitoring involves systematically collecting and analyzing data on ecological conditions to assess and track changes in the range, structure, function and species composition of habitats over time. This process uses methods like camera trapping and passive acoustic monitoring to gather information for effective conservation, restoration and biodiversity net gain (BNG) management plans, which ensure the long-term health (integrity) and sustainability of ecosystems.

Purposes of habitat monitoring include:

- Track trends: To observe changes in habitat extent, size, and location over time.
- Assess condition: To measure habitat health by quantifying its biotic (living), abiotic (non-living), and landscape characteristics.
- Evaluate effectiveness: To determine the impact of conservation actions and habitat management techniques.
- Inform adaptive management: To provide data that guides adjustments to conservation efforts for improved outcomes.
- Achieve biodiversity net gain (BNG): To create, enhance, and ensure the maintenance of habitats for a minimum of 30 years post-development.
- Obtain biodiversity credits for nature.

Common monitoring methods are:

- **Automated monitoring**: Using devices like camera traps or acoustic sensors to collect data on wildlife presence and activity, which can then be analyzed using artificial intelligence.
- **Remote sensing**: Employing satellite or aerial imagery to map land cover and track changes in habitat over large areas.
- **Field surveys**: Conducting systematic observations and measurements of vegetation, species, and environmental conditions at specific sites.
- Citizen science: Engaging volunteers to collect data using consistent protocols or habitat survey forms, although site selection bias may need to be addressed.

Key components of a monitoring plan are as follows:

- **Data collection**: Gathering information on biotic, abiotic and landscape characteristics.
- **Data analysis**: Processing the collected data to identify trends, assess conditions and measure the success of management interventions.
- **Adaptive management**: Implementing changes to management strategies based on monitoring results to improve conservation effectiveness.
- **Reporting**: Communicating findings to decision makers, partners and the public.

Figure 3.3 presents a combined process of habitat planning and AI applications for ecological prediction and optimization modeling, with a focus on optimum developments with mitigation measures.

3.2.8 Step 8: Climate Tech Credits

'Climate tech credits' can refer to carbon credits (offsets generated by projects reducing greenhouse gases) or tax credits (like the US federal Investment Tax Credits (ITCs) and Production Tax Credits (PTCs)) used to incentivize climate technology development. Carbon credits allow companies to offset their own emissions by funding projects that reduce or remove greenhouse gases, while tax credits provide

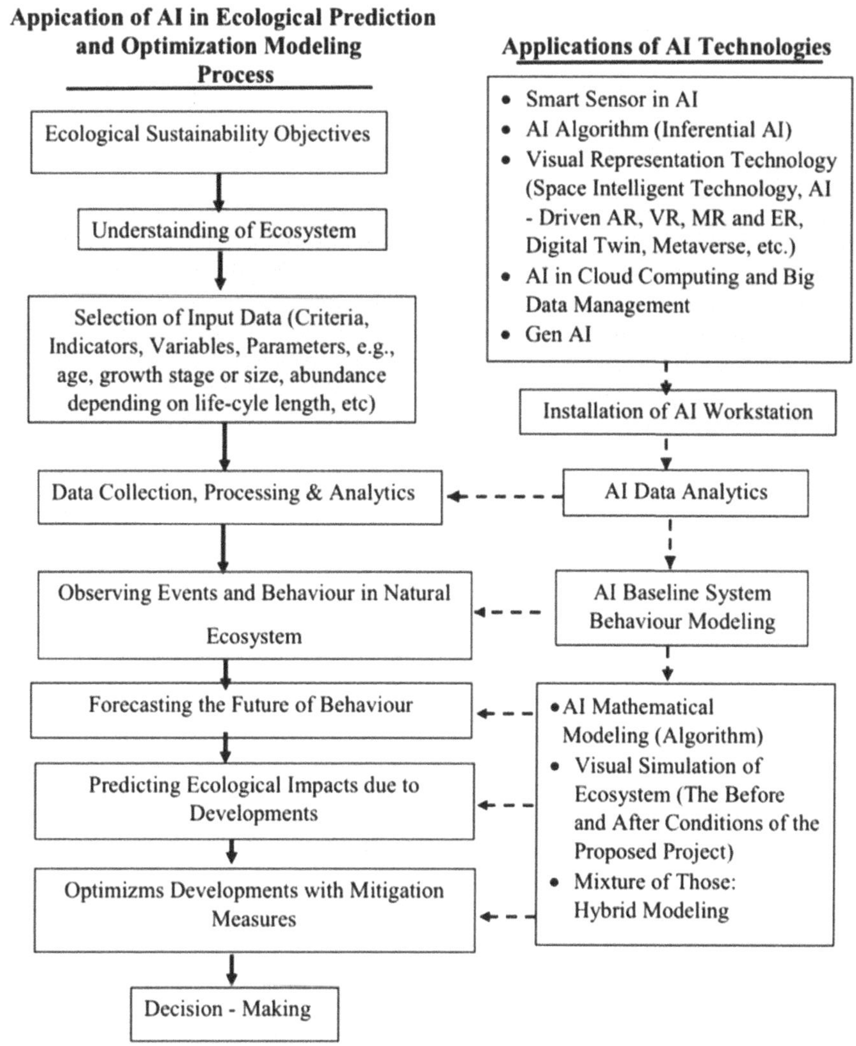

Fig. 3.3 Combined process of habitat planning and AI applications

financial benefits for investing in or producing renewable energy and other green technologies.

AI can be used in the future of carbon-financing tokenization for more accurate carbon accounting. In the current AI era, tokenization has become indispensable for Large Language Models (LLMs). With the surge of blockchain and cryptocurrency, tokenization took another leap, representing real-world assets digitally. Tokenization is an incredibly adaptable technology, and its significance has increased over time across diverse sectors (Fig. 3.4).

Fig. 3.4 Hana Bank AI finance card for customer asset management

3.3 Summary of AI Prediction and Optimization Methods

Table 3.4 can be used to identify potential AI prediction methods of selected themes. A list of AI optimization is shown for the similar themes in Table 3.5.

3.4 An Application

Table 3.6 shows how the Kitakyushu carbon neutrality plan could be improved with the AI technology described in the framework above, although more research is needed to optimize it.

The range of potential AI solutions described in Table 3.6 can be built over the combined image of energy and AI solutions. AI image can add another layer of data to the Fig. 3.5 to build the second layer solutions by AI means.

Table 3.4 Summary of AI prediction methods for selected themes: an illustration

AI prediction methods	Land use allocation	Sustainability cost–benefit analysis	Climate impact assessment	Habitat suitability analysis	Water quality analysis	Carbon neutrality city planning
1. Supervised learning methods *Regression models* Linear regression Polynomial regression XGBoost (Extreme gradient boosting) *Classification models* Logistic regression Decision tree Random forest Support vector machines (svMs) *Deep learning methods* Neutral networks Artificial neutral networks (ANNS) Recurrent neutral networks (RNNS) Long short-term memory (LSTM) networks Convolutional neutral networks (CNNs) Transformer models *Time series forecasting methods* ARIMA (Autoregressive integrated moving average) ETS (error, trend, seasonality) AI-driven time series forecasting	x	x	x	x	x	

(continued)

Table 3.4 (continued)

AI prediction methods	Land use allocation	Sustainability cost–benefit analysis	Climate impact assessment	Habitat suitability analysis	Water quality analysis	Carbon neutrality city planning
2. Unsupervised learning methods K-means clustering Anomaly detection						x

Note Climate tech credits and human behavior analysis can be included in the list of climate actions

Table 3.5 Summary of AI optimization methods for selected themes: an illustration

AI optimization methods	Land use allocation	Sustainability cost–benefit analysis	Climate impact assessment	Habitat suitability analysis	Water quality analysis	Carbon neutrality city planning	Climate tech credits	Human behavior analysis
1. Model optimization techniques Hyper-parameter tuning Model pruning • Quantization Knowledge learning Early stopping	x			x				x
2. Data-related optimization techniques Data preprocessing Data augmentation Data distillation		x			x		x	

(continued)

Table 3.5 (continued)

AI optimization methods	Land use allocation	Sustainability cost–benefit analysis	Climate impact assessment	Habitat suitability analysis	Water quality analysis	Carbon neutrality city planning	Climate tech credits	Human behavior analysis
3. Algorithmic and systematic optimization Gradient descent Efficient algorithms Data pipeline optimization Hardware compilation			x					
4. Broader optimization approaches Automated machine learning (AutoML) Explainable AI(XAI)						x		

Table 3.6 An application of the AI framework: the case of Kitakyushu City action plan for global warming countermeasures (2021–2030), Japan. Empowering cities to accelerate carbon neutrality action plan with AI (Kitakyushu City 2014)

Content of the action plan	Descriptions	Suggestions for the AI framework
1. Background and purpose	• SDGs–future city • Paris agreement and IPCC Special report on global warming of 1.5 °C Japanese government's declaration on achieving carbon neutrality by 2050	Framework 1. SDGs (SDG 13) Use of SDG 13, sustainable development targets and indicators Understanding the nexus between sustainability and AI cities
2. Basic concept	Grand design on the world capital of sustainable development: city with boundless 'true wealth and prosperity'	Framework 2. AI city climate agents for climate solutions Development of AI agents based on sustainable development targets and indicators Selection of sectors for single AI agents and multi-AI agents
3. Kitakyushu today and future estimates	• 15.77 million tonnes (2017) Estimated emissions in fiscal 2030 will be about the same as they are today (16.04 million tonnes), if no additional reduction measures are taken in the future	• Frameworks 3, 4, 5, and 6 To deploy AI sensors to collect data for CO_2 emissions inventory To consider carbon sequestration potential To manage and improve urban AI infrastructure, from traffic flow to resource management To establish urban big data platform with useful information from related open source software like Hadoop, Spark, and Mongo DB which are global cloud database services To use cloud-computing platforms such as Microsoft Azure and Amazon Web Services for preprocessing, managing and mining datasets that are continuously generated in cities To support AI computing infrastructure To predict BAU scenarios with AI technology
4. GHG reduction targets with additional reduction measures	Backcasting approach Normative approach based on future sustainability goals, targets and indicators (abstract world) Determination of reduction targets for fiscal 2030 (9.8 million tons) Selection of specific reduction measures To start with goals and end-year setting	Framework 7: AI urban data analytics—prediction and optimization (1) Forecasting approach (problem-solving approach to real world) based on learning data (2) To use AI deep learning workflow for prediction of emissions reduction potential of mitigation measures and optimization of additional reduction measures for target setting To start with methods of collecting learning data

(continued)

Table 3.6 (continued)

Content of the action plan	Descriptions	Suggestions for the AI framework
5. Aspiring to become a decarbonizing society by 2050 6. Mitigation measures by sector (towards 2030)	5.1 Kitakyushu's vision as a decarbonized society Development of Kitakyushu model and extending this model to other areas in Japan and abroad To decarbonize energy, promote innovation, transform lifestyle, and develop a resilient city capable of adapting to climate change 5.2 Formulating the Kitakyushu city green growth strategy Formulation of a sectoral plan to strategically secure sources of decarbonized energy and accelerate innovation in the future Mitigation measures by sector/field Methods to calculate reductions: (1) (BAU)–(2) reductions (already planned)–(3) reductions (from additional actions) Kitakyushu city's strategic vision on the future of energy Eco-town Energy grid system (hydrogen, renewable energy, power generation facility (waste plant), energy storage system, etc.) Key city policies Provide subsidies for the introduction of next-generation vehicles Promote the use of public transportation (formation of intercity BRT etc.) Promote the use of bicycles (Kitakyushu Bicycle Utilization Plan) Promote eco-driving and reduced use of private vehicles Use FCVs and EVs to secure emergency power sources during disasters Disseminate information on the use of new mobility systems	AI BAU scenarios AI energy grid system management AI climate tech credits
7. Green government initiatives	Setting reduction targets for GHG from city operations by 60% from FY 2013 levels Generating power from 100% renewable energy sources in public facilities as a leading model for the creation of a decarbonized society	

(continued)

Table 3.6 (continued)

Content of the action plan	Descriptions	Suggestions for the AI framework
8. Contributions to decarbonization efforts around the world	Creation of a 'virtuous cycle' between the environment and economy Taking on a leading role with the Asian Center for Low Carbon Society (established in 2010)	
9. Adaptation to climate change impacts (adaptation measures)	To promote adaptation measures with reference to the results of climate change impact assessment by the Japanese government	Applications of AI in climate impact assessment
10. Promotion of plan	To establish "progress management indicators" to ensure that progress is managed effectively To develop a follow-up system	Use of sustainability indicators
11. Climate tech credits	Not applicable for Kitakyushu City	Cities are putting AI in carbon financing initiatives' hands: To tokenize climate tech credits: Crypto ventures To explore LCA approach to applications of AI in every step of carbon FinTech such as data sensing, storage and data analytics, trading and monitoring for both financing and climate tech credits To broaden climate tech credits to incentivize climate technology development Technology-based solutions (emissions reduction) Nature-based solutions (removal by sinks) To improve energy efficiency: Incentives to save energy in energy and water grid system, buildings, transportation, and industrial processes, and then to reduce GHG emissions Energy demand predictive analytics Optimization of operation To promote additionality of all technology at the city level

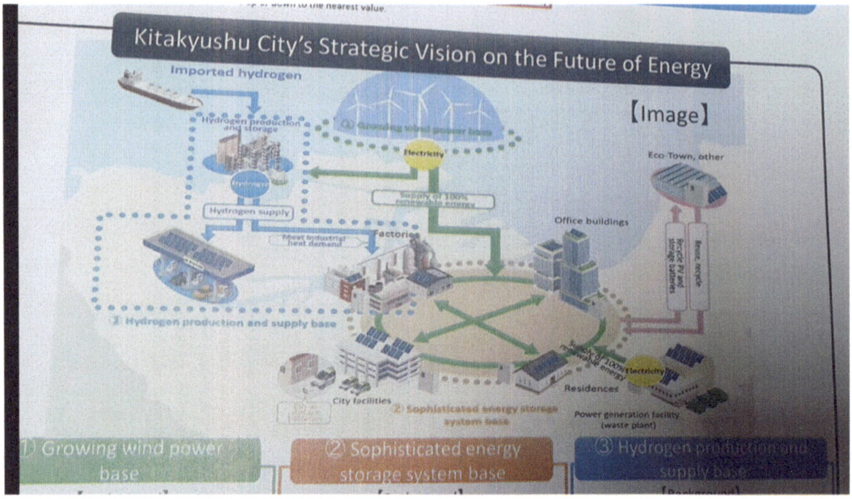

Fig. 3.5 Kitakyushu City's strategic vision on the future of energy. *Source* Kitakyushu City (2024)

References

Kitakyushu City (2024) Kitakyushu action plan for global warming countermeasures (2021–2030), Overview, Kitakyushu City Government, Japan

Lee C (1974) Models in planning: an introduction to the use of quantitative models in planning. Pergamon Press, Oxford, p 1

Treweek J (1999) Ecological impact assessment. Blackwell Science, Oxford, p 16

World Resources Institute (2025) Insights, 11 September

Chapter 4
AI Agents

Kwi-Gon Kim

Abstract AI city agents have great potential for climate solutions. Multi-AI agents are more holistic than multi-tiered assessment based on a map overlay method. The sustainability indicator approach to optimization involves using specific metrics to guide decision making towards more sustainable outcomes. This approach integrates environmental, social and economic considerations into the optimization process, ensuring that solutions are not only efficient but also contribute to long-term well-being and resource conservation. This chapter describes relationships between climate solutions and AI agents by using eco-cultural tourism, different layers of functions for AI workloads, categorization of AI agents rebuilding optimism in the age of climate crisis, a framework for multi-AI agents for model innovation and methodologies, and tools of AI agents for climate solutions.

Keywords AI agent · Digital workforce · Software system · Overlay · AI workload · Layers of AI functions · Machine learning flow · Ceph storage · Multi-AI agent · AI ecological model · Land allocation model

4.1 Relationship Between Climate Solutions and AI Agents

AI agents tackle climate solutions by autonomously monitoring and analyzing data, optimizing energy use and grid management, improving climate and disaster forecasting, and supporting sustainable practices. These autonomous software systems enhance scalability and speed in addressing complex climate challenges, acting as a digital workforce to automate tasks, provide real-time insights and help cities make data-driven decisions for climate action.

Each of the AI agents needs to be analyzed separately prior to coming together to form interactive multi-AI agents. Figure 4.1 shows a platform model for AI agents in a connected AI world (Table 4.1).

K.-G. Kim (✉)
Seoul National University, Seoul, South Korea
e-mail: kwigon@snu.ac.kr

K.-G. Kim (ed.), *AI Transformation of Cities for Climate Solutions*,
https://doi.org/10.1007/978-3-032-21940-4_4

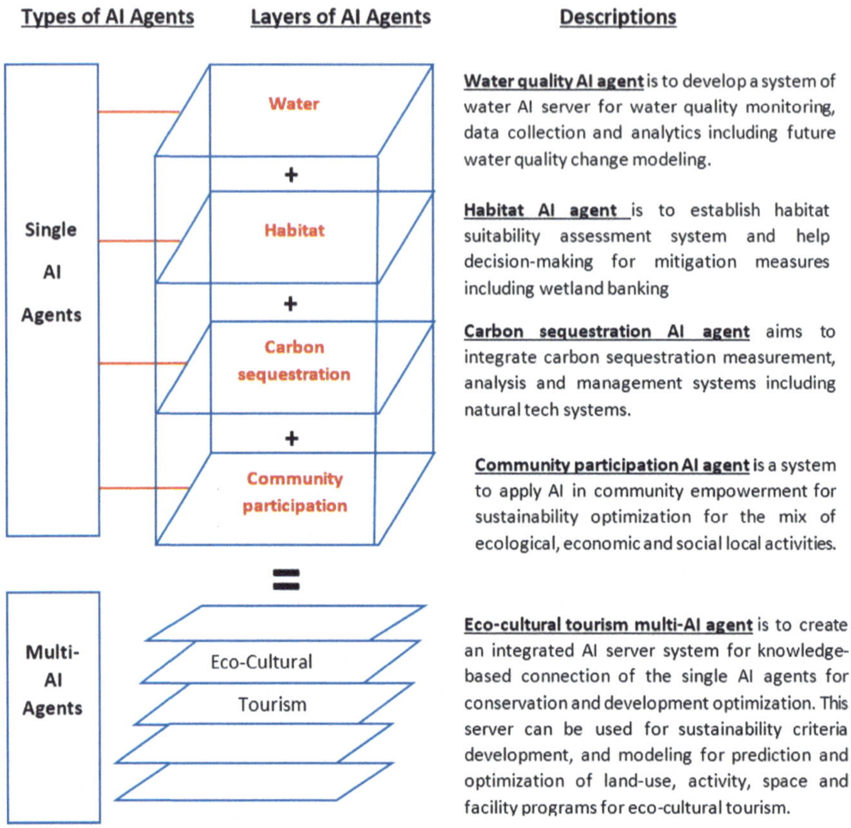

Fig. 4.1 Overlay of AI agents: AI agent components with 3D digital mapping

The following table shows five use cases of AI agents in eco-cultural tourism planning, development and operation.

4.2 Different Layers of Functions for AI Workloads

The author met with a German AI expert (Benedikt Burk, Head of Professional Services, CLYSO, Munich, Germany) to discuss the KOICA PCP in Seoul. He has kindly prepared an important and concise paper 'Overlay of Functions for AI Application,' from data sensing to operation and management (Fig. 4.2). The key process is to integrate connected data with AI to predict and optimize the project with the necessary equipment identified in this book.

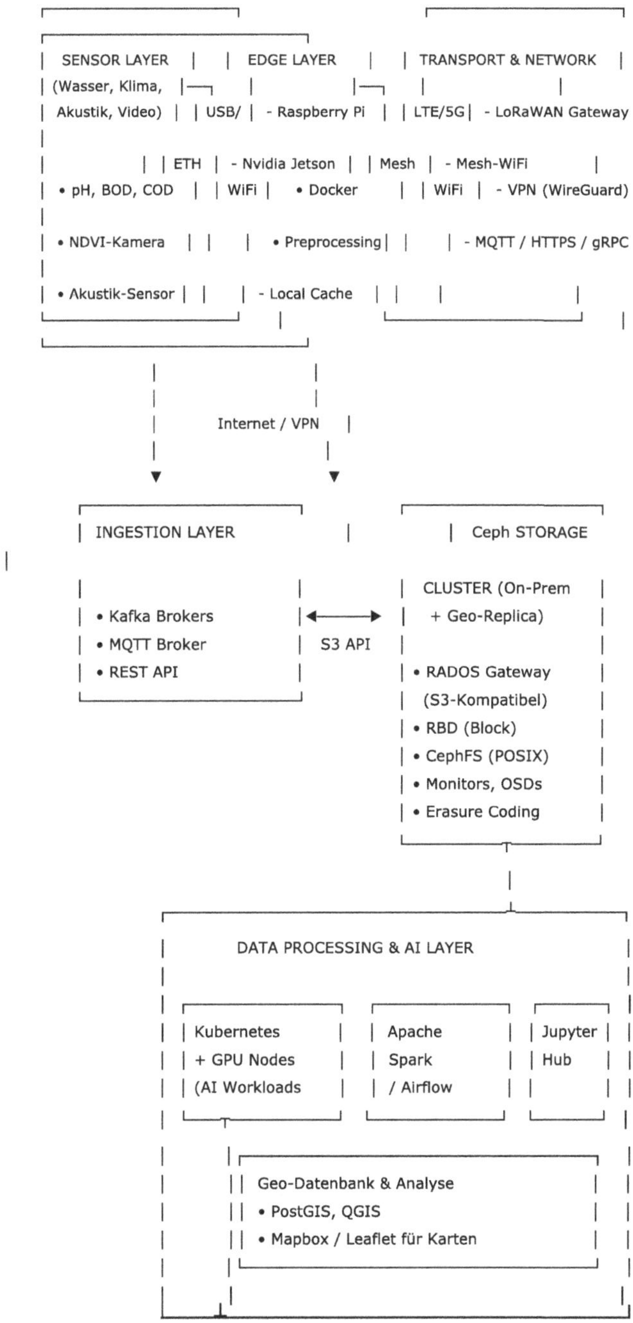

Fig. 4.2 Different layers of functions for AI workloads

Fig. 4.2 (continued)

```
        |
        ▼
┌─────────────────────────────────┐
|  APPLICATION & CLIENT           |
|                                 |
| • React / Vue Web-Portal        |
| • Grafana / Superset Dashboards |
| • Rasa / LangChain Chatbots     |
| • Flutter / React Native Apps   |
| • AR/VR (Cesium, Unity)         |
└─────────────────────────────────┘
        |
        ▼
┌─────────────────────────────────┐
|  OPERATIONS & SECURITY LAYER    |
|                                 |
| • Prometheus + Grafana          |
| • ELK / Loki Logging            |
| • Keycloak IAM                  |
| • Vault Secrets Management      |
| • CI/CD (ArgoCD, GitLab CI)     |
| • Vulnerability Scans (Trivy)   |
└─────────────────────────────────┘
```

The architecture stack—the layer functions—of Fig. 4.2 is summarized below:

(1) SENSOR LAYER

- What happens here?

This layer captures the physical world using various environmental sensors and cameras.

Key functions:

Monitor water quality (pH, BOD, COD, etc.)

Detect birds (e.g., storks) via acoustic and camera sensors

Record climate, air quality, and vegetation health (e.g., NDVI)

Count visitors using video feeds or sensors

(2) EDGE LAYER

- What happens here?

Initial data processing takes place at the edge, close to the sensors.

Key functions:

Preprocess data (filtering, compression, caching)

Perform real-time AI inference (e.g. bird detected?)

Buffer data during network outages

Run containerized apps (e.g. Docker on Jetson or Raspberry Pi)

(3) TRANSPORT AND NETWORK LAYER

- What happens here?

This layer moves data securely and reliably from sensors to central systems.

Key functions:

Communicate via LoRaWAN, LTE/5G, or mesh Wi-Fi

Use protocols like MQTT, HTTPS, WebSockets

Ensure security with VPNs (e.g. WireGuard)

(4) INGESTION LAYER

- What happens here?

This is the entry point for all incoming data streams.

Key functions:

Receive data via brokers (Kafka, MQTT)

Expose REST APIs for external apps

Timestamp and tag data for downstream processing

Forward data to storage and processing systems

(5) STORAGE LAYER (Ceph) - > CLYSO

- What happens here?

All project data is stored securely and scalably in a Ceph-based system.

Key functions:

Object storage (Ceph RADOS Gateway) for images, logs, sensor data

Block storage (Ceph RBD) for AI workloads and VM disks

Shared file system (CephFS) for collaboration

High durability via erasure coding and replication

Optional cloud sync for off-site backup or hybrid use

(6) DATA PROCESSING & AI LAYER

- What happens here?

This is the project's brain—where AI, analytics and modeling take place.

Key functions:

Train and deploy AI models (e.g., stork detection, visitor prediction)

Run batch and real-time analytics (Spark, Jupyter)

Perform GIS analysis with PostGIS, QGIS

Forecast environmental and tourism-related patterns

Use MLops tools (e.g. ML (machine learning) flow) for model management

(7) APPLICATION & CLIENT LAYER

- What happens here?

User interfaces for citizens, tourists, researchers, and local government.

Key functions include:

Web and mobile apps for tourists and stakeholders

Dashboards to display environmental metrics and AI predictions

Chatbots for eco-tourism information and local engagement

AR/VR tools for immersive tourism experiences

(8) OPERATIONS & SECURITY LAYER

- What happens here?

Ensures system uptime, reliability, and security.

Key functions:

System monitoring (Prometheus, Grafana)

Logging and debugging (ELK Stack, Loki)

Identity and access management (Keycloak, OIDC)

Secrets and credential management (Vault)

DevOps pipelines (ArgoCD, GitLab CI/CD)

Security scanning and vulnerability management (e.g. Trivy)

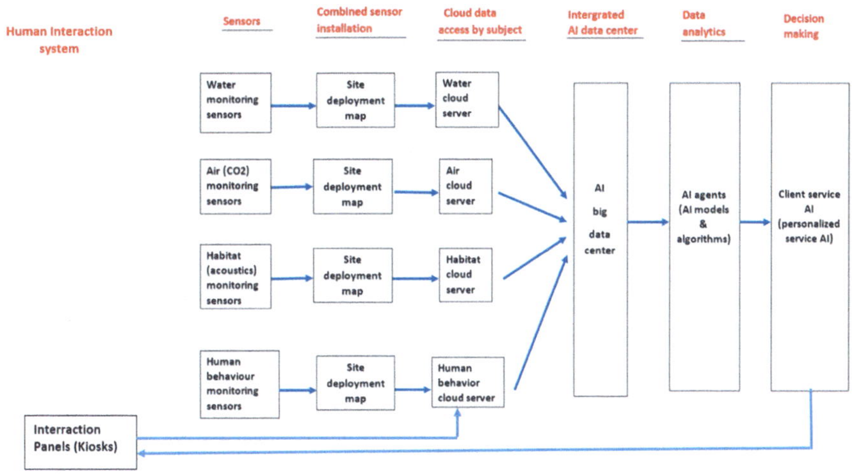

Fig. 4.3 Integrated AI schema

- How All Layers Work Together

Sensors capture the physical environment → edge devices preprocess the data → data is securely transmitted to Ceph storage → AI models process and analyze it → insights are shared through applications and dashboards → all backed by secure, observable and maintainable infrastructure.

The author has designed a series of frameworks for AI-integrated approaches in which AI will be used to enhance climate-responsive planning by precisely targeting an AI schema (Fig. 4.3).

Table 4.2 gives the readers an additional explanation for Fig. 4.2 in terms of stage of AI agent, AI devices and key functions.

These processes will be used for developing optimization algorithms in the next section.

4.3 Categorization of AI Agents Rebuilding Optimism in the Age of Climate Crisis

AI agents are broadly categorized into five main types: simple reflex, model-based, goal-based, utility-based, and learning agents. These categories are based on how the agent perceives its environment, makes decisions and learns. On the other hand, multi-AI systems consist of multiple agents working together to achieve a common goal, combining the strengths of different agent types. These computer systems are like digital teams, where multiple independent players—called agents—work together (or sometimes compete) to get things done.

Table 4.1 Five use cases of AI agents in eco-cultural tourism planning, development and operation

Use case	Description
Single AI agent 1: Integrated water quality management system	• Gathers information, compiles reports and visualize metrics, helping analysts work more efficiently • Streamlines inventory and decision making • Optimizes water-management practices • Enables collaborative workflows for water-quality inventory and management
Single AI agent 2: Stork habitat evaluation and management	• Predicts change in stork habitats • Drafts optimum habitat areas to optimize stork performance (mating, feeding, daytime activities, hiding and sleeping) based on index data-habitat evaluation
Single AI agent 3: Carbon sink and storage	• Predicts future carbon sequestration potential • Generates targeted carbon banking markets • Classifies and monitors sensitive data to support regulatory carbon-trading compliance and prevent breaches
Single AI agent 4: Local participation	• Automates local support by drafting responses, analyzing past experiences, and personalizing replies • Generates personalized follow-up emails • Designs adaptive, personalized learning paths based on local skills and performance • Develops local people's chatbot
Multi-agent generative AI for multiple teams: Combined Single AI agents of 1,2,3 and 4 above	• Creates and manages Multi-AI agents tailored to the unique needs of different teams, all from single platform • Forecasts eco-cultural tourism demand • Helps legal teams scan contracts for risks, missing clauses and discrepancies, speeding up the review process • Predicts revenue, expenses and cash flow, using historical data and real-time financial data • Automates coding, debugging and collaboration tasks to boost developer productivity • Creates targeted, data-driven campaign using visitor behavior and purchase history • Offers round-the-clock assistance via chat, email and social performs, using Perplexity and automation • Drafts content, manages campaigns and analyzes performance to optimize messaging, based on data-driven insights

Table 4.2 Devices and functions for AI workloads towards convergence of theory and technology for sustainability optimization: creation of eco-cultural tourism site with data and AI

Stage of AI agent	AI devices (hardware)	Key functions (software)
1. AI sensing	Water quality monitoring • Sensors for pH, BOD, COD etc Carbon modeling and carbon Sequestration potential • Harmful gas sensors • CO_2 sensors • Meteorological sensors (temperature, humidity, wind etc.) • Solar-powered chlorophyll meters • Photosynthesis sensors Habitat suitability evaluation • Acoustic and camera sensors • Wireless camera sensors for animal movement • Sensors for biomass growth rate Community participation (human behavior interaction) • AI apps • AI counts for visitors and bikes • Basic monitoring dashboard • Edge devices for cameras	• To monitor water quality (pH, BOD, COD,etc.) • To predict water quality change • To record climate, air quality, and vegetation health (e.g. NDVI: normalized difference vegetation index) • To collect data for calculating carbon sequestration potential • To track and analyze carbon footprint • To detect birds (e.g. storks)via acoustic and camera sensors • To collect data for calculating habitat suitability indices (multicriteria techniques) • To count visitors using video feeds or sensors • To analyze user behavior • To monitor AI-powered urban behavior (tourism behavior) • To integrate with existing surveillance feeds
2. Initial data processing	• Rasberry Pi • Nvidia Jetson Docker Preprocessing • Local Cache	Initial data processing takes place at the edge, close to the sensors • To preprocess data (filtering, compression, caching) • To buffer data during network outages • To run containerized apps (e.g. Docker on Jetson or Rasberry Pi)

(continued)

Table 4.2 (continued)

Stage of AI agent	AI devices (hardware)	Key functions (software)
3. Data transport and network	• LoRaWAN Gateway • Mesh-WiFi • VPN (WireGuard) • MQTT/HTTPS/ gRPC	This layer moves data securely and reliably from sensors to central systems • To communicate via LoRaWAN, LTE/5G, or mesh WiFi • To use protocols like MOTT, HTTPS, WebSockets • To ensure security with VPNs (e.g. WireGuard)
4. Ingestion	• Kafka Brokers • MOTT Broker • REST API	This is the entry point for all incoming data streams • To receive data brokers (Kafka, MOTT) • Expose REST APIs for external apps • To timestamp and tag data for downstream processing • To forward data to storage and processing systems
5. Data storage	Ceph STORAGE CLUSTER (On-PRem + Geo-Replica) • RADOS Gateway • S3-Kompatibel • RBD (Block) • CephFS (POSIX) • Monitors, OSDs = Erasure Coding	All project data are stored securely and scalably in a Ceph-based system • Object storage (Ceph RADOS Gateway) for images, logs, sensor data • Block storage (Ceph RBD) for AI workloads and VM disks • Shared file system (CephFS) for collaboration • High durability via erasure coding and replication • Optional cloud sync for off-site back-up or hybrid use
6. Data processing and AI	• Kubernetes + GPU Nodes (AI Workloads) • Apache Spark/ Airflow • Jupyter Hub • Geo-Datenbank and Analyse • PostGIS, QGIS • Mapbox/Leaflet for Karten	This is the project's 'brain' where AI, analytics, and modeling take place • Train and deploy AI models and algorithms (e.g., stork detection, habitat optimization, water-quality change prediction, visitor prediction, carbon-sequestration potential, human interaction system, energy efficiency and ecological optimization) • Run batch and real-time analytics (Spark, Jupyter) • Perform GIS analysis with PostGIS, QGIS • Forecast climate, environmental and tourism-related patterns and impacts • Use MLops tools (e.g. MLflow) for model management

(continued)

Table 4.2 (continued)

Stage of AI agent	AI devices (hardware)	Key functions (software)
7. Application and client service AI	• React/Vue Web-Portal • Grafana/Superset Dashboards • Rasa/LangChain Chatbots • Flutter/React Native Apps • AR/VR (Cesium, Unity)	User interfaces for citizens, tourists, researchers and local governments • Web and mobile apps for tourists and stakeholders • Dashboards to display environmental metrics, and AI predictions and optimization • Chatbots for eco-tourism information and local engagement • AR/VR tools for immersive tourism experiences
8. Operation and security	• Prometheus + Grafana • ELK/Loki Logging • Keycloak IAM • Vault Secrets Management • CI/CD (ArgoCD, GitLab CI) • Vulnerability Scans (Trivy)	Ensures system uptime, reliability and security • System monitoring (Prometheus, Grafana) • Logging and debugging (ELK Stack, Loki) • Identity and access management (Keycloak, OIDC) • Secrets and credential management (Vault) • DevOps pipelines (ArgoCD, GitLab CI/CD) • Security scanning and vunerability management (e.g. Trivy)

Some multi-agent systems are inspired by nature, mimicking how ants work together to build complex colonies or how birds flock in perfect synchronization (ChatGPT 2025).

From virtual worlds to real-life applications, the case study of multi-AI agents in the Song Dam Lake project uses four single AI agents including water, animal habitat, carbon sequestration and community participation AI agents (depicted in Fig. 4.1) to achieve common goals of sustainable development.

4.4 Framework for Multi-AI Agents for Model Innovation

Multi-agent vs single-agent model comparisons also reveal a notable difference in task management. Single-agent models work on a single thread, limiting them to one task at a time, which can slow down responses in environments that require quick multitasking. Multi agent systems, however, operate in parallel, allowing various agents to address different tasks simultaneously.

Collaboration within multi-agent systems in artificial intelligence opens up capabilities by bringing together diverse skill sets. Where single-agent setups might struggle with complex scenarios, multi-agent frameworks combine the strengths of specialized agents to form well-rounded solutions.

The structure of a multi-agent system is designed for agents to communicate and work toward shared objectives. Each agent is equipped with unique methods

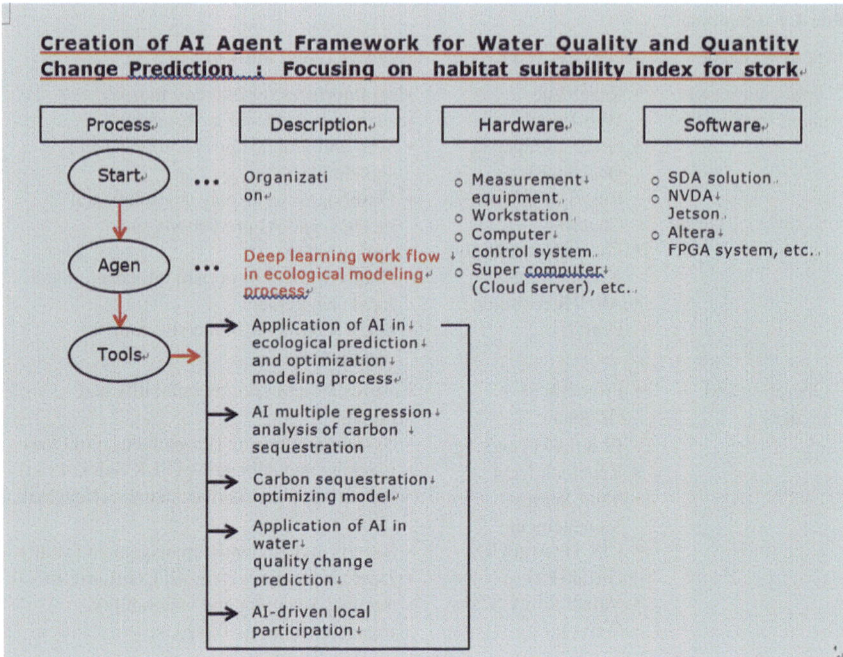

Fig. 4.4 Structure of a multi-AI agent system

for solving problems and can interact with others within the system. This multi-agent architecture facilitates dynamic, cooperative decision-making, allowing for more comprehensive handling of complex and ever-changing tasks (Refer to "3.2 Different layers of functions for AI workloads").

Figure 4.4 shows the structure of a multi-AI agent system while Table 4.3 describes collaboration within multi-agent systems in AI, based on Fig. 4.4.

4.5 Methodologies of AI Agents: AI-Powered Project Activities to Create Profiles

The methodologies of AI agents for different project activities showcase potential use of digital sensors, collection of data, storage of data, analysis of data and intelligent algorithms. The intelligent algorithms for urban planning and execution include:

- Deep learning and optimization
- Graph and regression models which can be used to interpolate, predict and forecast unobserved values based on given data
- Human-machine learning systems which are needed to organically integrate the knowledge from both sides to tackle an urban challenge.

Photo 4.1 Photo of storks presented by Vice Chair Mr. Hau in memory of Prof. Kim's Keynote Speech on AI Eco-Cultural Tourism at the International Conference held in Tam Ky City, 2024

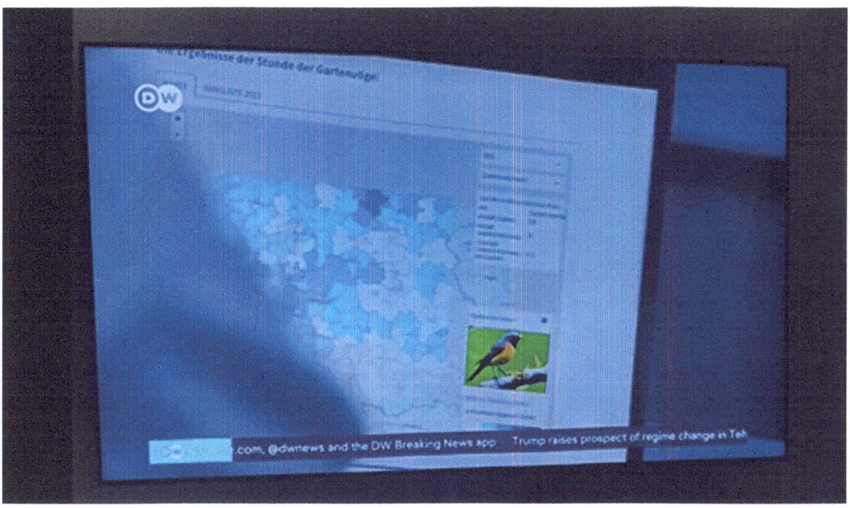

Photo 4.2 Deployment of acoustic wave sensors throughout Germany. *Source* DW, 23 June 2025

The following sections present some promising methodologies of AI agents selected from the city components in Table 4.4.

Photo 4.3 Installation of acoustic wave sensors on trees: An AI developer and an ornithologist. *Source* DW, 23 June 2025

Photo 4.4 German ornithologist explaining AI technology for acoustic wave sensing. *Source* DW, 23 June 2025

(1) Agent-Based Ecological Model

Agent-based ecological models rely on high-quality, real-world data to accurately represent ecological systems and produce meaningful results. For building a digital

Photo 4.5 Use of computer servers to receive electrical signals from sensors. *Source* DW, 23 June 2025

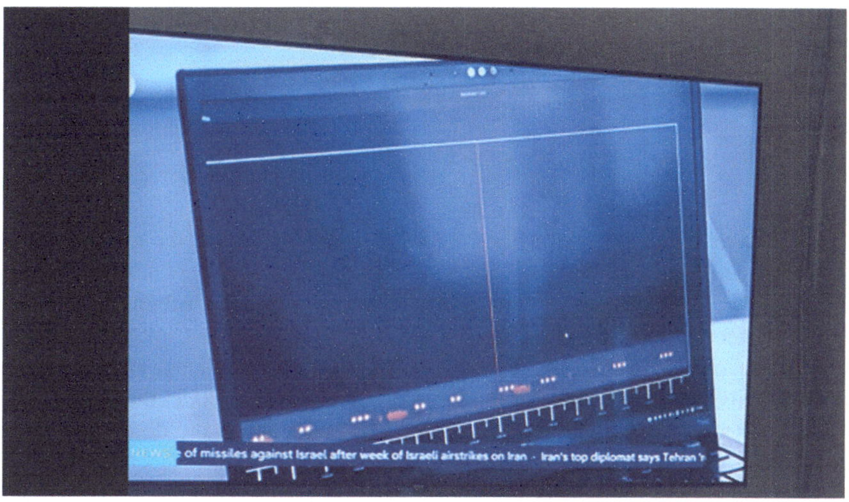

Photo 4.6 Interpretation of electrical signals to identify bird species with sound waves displayed on the computer screen. Different bird species have different electrical signals (sounds). *Source* DW, 23 June 2025

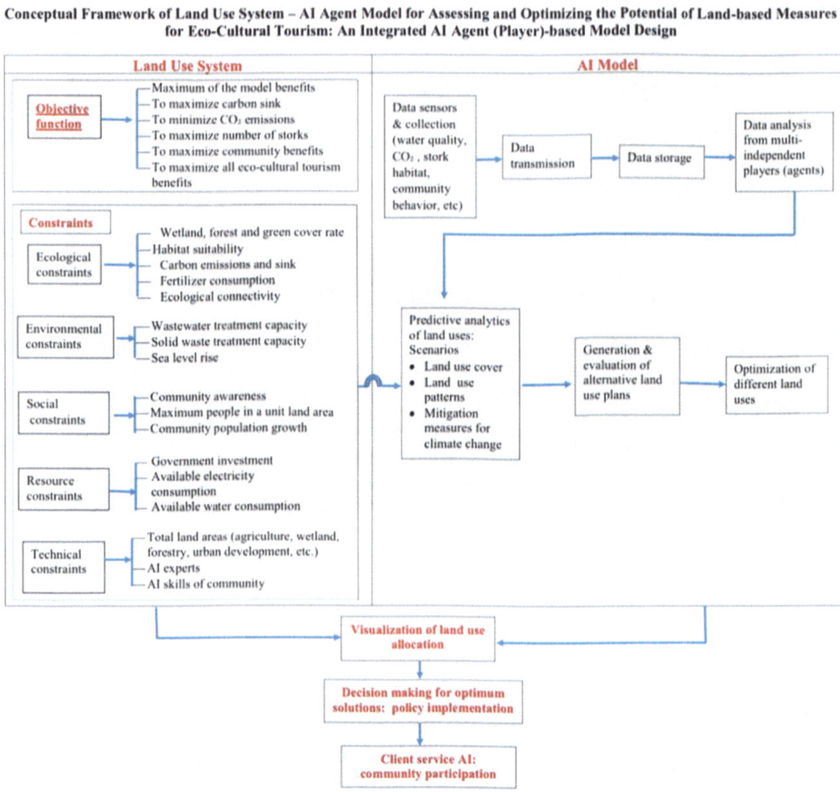

Fig. 4.5 Multi-AI agent model for eco-cultural tourism planning

twin of nature, the more detailed and frequent our observations, the more realistic our virtual ecosystem becomes.

Modern ecological research has transformed how we collect environmental data. Advanced sensor networks and field tools now generate vast amounts of high-fidelity data, capturing everything from animal movements to changes in vegetation patterns. This wealth of information allows models to better reflect the complexity of natural systems. The details of the model and its execution are described in the next section.

(2) Stork Habitat Using Habitat Suitability Index (HSI): AI Biodiversity Optimization Model

Acoustic sensors are devices designed to detect, measure, and analyze sound waves in various mediums, including air, water, or solids. These sensors convert sound waves into electrical signals, allowing for the detection and interpretation of acoustic signals.

Table 4.3 Collaboration within multi-agent systems in AI

City components (unknown AI application areas: AIX (AI transformation))	Sustainability indicators (for assessing benefits of AI)	AI agent framework (optimized expert models that evolved from foundation models)	AI tools (optimized expert tools that are 'advanced and individually customized' including text information, voice recognition, LLM, image generation, video generation, data search and data farming): Intelligent data analytic algorithm)	Decision making
Description	Sustainability is needed to develop and evaluate structure of AI agent and tools Sustainability indicators and criteria provide common goals of creating climate-smart and sustainable cities and describes, in detail, the value criteria used to characterize AI Agents and tools for sustainability transformation	An AI agent is the system or program designed to perform the specific work for prediction and optimization	55 AI tools (software) across 25 different categories are available from 'Synthesia, The Best AI Tools by Category'	
Land use and urban form		AI land use agent: Land Use Plan/Design Model determining the area of land, by major land use category to be developed and conserved in each zone		

(continued)

Table 4.3 (continued)

Habitat evaluation		AI habitat suitability agent: Habitat Suitability Index Model comparing habitat conditions with optimum habitat conditions for the same evaluation species (e.g. stork)		
Calculation of carbon sequestration potential		AI carbon agent		
LCA/ consumption-based green-house gases (GHG) emission accounting		AI GHG accounting agent		
Community participation		AI local participation agent		
Eco-cultural tourism		AI eco-cultural tourism agent		
CO_2 emission prediction system		AI CO_2 emission prediction agent		
Energy grid		Multi-AI energy grid agent: Predicting energy generation and consumption		
Blue-green infrastructure		AI blue-green infrastructure agent: Creating the network		
Local participation		AI local participation agent: Empowering local people		

(continued)

Table 4.3 (continued)

City grid system		Multi-AI city grid agent: Improving efficiency		
Cars moving		AI-cars moving agent: Creating efficient mobility systems		
Carbon financing		AI carbon financing agent: Supporting climate action		
Climate modeling		AI climate modeling agent: Forecasting weather		
Personal footprint		AI footprint agent: Improving personal behavior		
Urban simulation of cities		AI simulation agent: Creating a new image		
Urban lifestyle		AI lifestyle agent: Changing lifestyle		
Individual consumer's behavior		AI consumer's behavior agent: Reducing GHG emissions		
Fashion		AI fashion agent: Using local materials		
Smart grid		AI smart grid agent: Using urban big data analytics		
Transportation (street lighting control)		AI transportation agent: Encouraging intelligent mass transportation		
Urban watershed management		AI watershed management agent: Creating integrated systems		

(continued)

Table 4.3 (continued)

Urban wetland park		AI wetland park agent: Creating new benefits		
Built environment		Generative AI agents: Encouraging an holistic approach in the built environment		

- Basic steps associated with detecting storks are:

(1) Description of plant community types and their geographical distribution, and species for each community type

(2) Deployment of acoustic wave sensors in the selected site.

(3) Conversion of sound waves into electrical signals

An agent-based ecological model can be used for detecting storks with AI technology.

(4) Interpretation of electrical signals

- Development and Application of AI Agents for Optimization of Animal Habitats: Integration of Big Data and AI Platform

(1) Collection and time series analysis of learning data for animal habitats

- Literature review
- Review of case studies
- Field surv ey
- Expert knowledge
- Use of AI servers (e.g. Google, IBM, Naver, Daum, etc.)

(2) Determination of optimum animal habitat criteria in the same environmental circumstances
(3) Testing of deep learning program to determine optimum habitats for animals in terms of mating (breeding), hiding, feeding, sleeping and daytime activities
(4) Application of the habitat AI agents and tools in the case study area: Integration of human and machine intelligence

3) Land Use Planning, Carbon Emissions and Sequestration

Planning models are defined as 'models which produce the plan which best satisfies the stated objectives, subject to the availability of resources and the constraints which are specified' (Lee 1974).

- Land Allocation Models and Their Applicability in Assessing Climate Change Mitigation Potential

Traditional land use models based on the map-overlay technique should be refined for multiple projects including land use patterns and land use change under different scenarios and with climate feedback by AI means.

Land allocation models highlight the interrelations between human activities and environmental changes, emphasizing the importance of sustainable land management to mitigate greenhouse gas emissions without compromising food security and ecosystem services.

The land allocation models are crucial for assessing climate change mitigation potential by simulating how different land uses (like agriculture, forestry and urban development) might change in response to climate policies, and how those changes affect greenhouse gas emissions and carbon sequestration. These models can help identify optimal land management strategies for reducing emissions, enhancing carbon sinks and achieving climate goals.

Several purposes are served by different land allocation models:

Spatial representation:

These models represent land cover and land use patterns across geographic areas, often at a high spatial resolution.

Land use change drivers:

These incorporate various drivers of land use change, including climate change impacts, economic factors (e.g. agricultural prices), population growth and policy interventions.

Land use allocation rules:

These use algorithms (e.g. linear programming, agent-based modeling, cellular automata) to determine how land is allocated to different uses under different scenarios.

Climate feedback:

This model simulates the impacts of land use changes on greenhouse gas emissions (both emissions from land use changes and emissions from other sectors influenced by land use) and on carbon sequestration in vegetation and soils.

Integration with other models: Land allocation models are often linked with other types of models, such as integrated assessment models (IAMs) and earth system models, to assess the broader impacts of land use changes on the climate system and other sectors.

In conclusion, land allocation models are valuable tools for assessing the potential of land-based measures for climate change mitigation. By simulating land use changes under different scenarios, they can inform the design of effective climate policies and help achieve a more sustainable and climate-resilient future.

- A Proposed Land Allocation Model for AI Eco-Cultural Tourism Development

The author has developed a land allocation model for AI eco-cultural tourism development taking into account objective function and constraints (Fig. 4.1). An important purpose of this model is to optimize different land uses which satisfy the stated objective function and the constraints which are specified in Fig. 4.5.

(4) Greenhouse Gas Emission Accounting from Automatic Real-Time Inventory-Based Data

Consumption/needs-based greenhouse gas (GHG) accounting can switch GHG accounting from annual to real-time or hourly accounting with personalized AI technology. This tool can be used to support climate impact assessment (CIA) over a 30-year time series in cities.

(5) Downscaled Climate Impact Assessment

In order to facilitate better planning for the impacts of climate change on proposed and existing assets, we need to develop and maintain downscaled climate impact assessments, based on the latest site observation and modeling data developed in this book.

The downscaled data and assessments could be incorporated into the system described in the Framework for City Climate Planning AI, and used to analyze proposed projects and land use decisions to improve the resiliency of new projects. The details of CIA are described in Chap. 5.

(6) Land Use, Community, Socio-Economy and Climate Integration

UrbanSim helps urban development stakeholders make smarter plans with less effort.

State-of-the-art urban simulation, 3D visualization and shared open data empowers users to explore, gain insights into, and develop and evaluate alternative plans to improve their communities. UrbanSim's platform incorporates interactions between land use, transportation, the economy and the environment to build better cities.

This platform can be extended to include other factors such as community patterns and lifestyle, and their relationship to land use and climate impact, using multi-AI agent.

(7) Sustainable Supply Chains

Sustainable supply chain analysis helps by utilizing AI for life cycle analyses (LCA) to measure the carbon footprint of consumer goods, and guide transitions to more sustainable and low-carbon products. Personalized client AI technology should

Table 4.4 Carbon sequestration potential by site

Site number	Variable (carbon sequestration potential (t C)/m2	Parameters					Hidden parameters			
		Leaf size (chlorophyll activity)	Biomass	Water health indicators			Future land use change	Mitigation measures	Future events	
				Water pollutants	Sediment	Water depth			Climate change	Flood risk
1										
2										
3										
.										
.										
.										

be used as it is more applicable for measuring the energy consumption of consumer goods and services and then reducing the carbon footprint of individual consumers.

(8) Carbon Sequestration

A plan evaluation approach can be utilized in carbon sequestration analysis. Table 4.4 shows parameters and variable for evaluating carbon sequestration potential by the selected site. AI agents can help the selection of the site that will maximize the potential of carbon sequestration.

(9) Other Innovative Subthemes of AI Agents

Deep Learning for Deforestation Detection
Predictive Modeling of Forest Growth
AI-Driven Carbon Sequestration Forecasting
Automated Carbon Flux Estimation
ML-Based Risk Detection in Carbon Sinks
Remote Sensing Data Analytics
Ground-Based Measurement Integration
Climate Data Utilization in AI Models
Soil Mapping with Machine Learning
AI-Enabled Topographic Analysis
Forest Health Monitoring with ML
Carbon Flux Prediction Models
AI for Ecosystem Resilience Analysis
ML-Based Forest Management Optimization
AI-Driven Adaptive Land Management
Carbon Sequestration Policy Modeling
Automated Carbon Offset Project Verification
Transparency in Carbon Offset Monitoring

ML for Environmental Integrity Assurance
AI in Sustainable Investment Evaluation

References

DW News (2025) Tomorrow Today, 23 June
Lee C (1974) Models in planning: an introduction to the use of quantitative models in planning. Pergamon Press, Oxford, p 113
ChatGPT (2025) website

Chapter 5
AI Modeling Methods, Tools and Techniques

Theoretical Combination of AI and City Climate Planning for Soft Computing Techniques

Kwi-Gon Kim, Catherine Atkin, Anna Grichting, Trieu Thy Hoa, and Seijoong Kwon

Abstract The climate software wave plays a central role in climate mitigation (reducing GHG emissions), climate adaptation (managing climate risks), climate intelligence (modeling, forecasting and decision support) and sustainability management (assessing climate impact, compliance, and disclosures). There is a growing acceptance that the use of AI for models can enhance planning by understanding and predicting the behavior of urban systems (structure and function) for climate-smart and sustainable solutions. In an era where artificial intelligence is rapidly transforming our relationship with planning, it is time to look beyond the dichotomy between natural and artificial intelligence, exploring the rich ecosystem of intelligence that shapes our cities. This chapter explores the role of models, AI prediction and optimization processes, and deep learning workflow towards theoretical combined methods of natural (human) and machine intelligence. It also describes

K.-G. Kim (✉)
Seoul National University, Seoul, South Korea
e-mail: kwigon@snu.ac.kr

C. Atkin
Stanford Codex Climate Data Policy Initiative, San Jose, USA
e-mail: atkin@codex.stanford.edu

A. Grichting
University of Vermont, Burlington, USA
e-mail: bordermeetings@hotmail.com

T. T. Hoa
Quang Nam University, Da Nang, Vietnam
e-mail: trieuthyhoa@gmail.com

S. Kwon
Ministry of Foreign Affairs, Seoul, South Korea
e-mail: sjkwon@mofa.go.kr

K.-G. Kim (ed.), *AI Transformation of Cities for Climate Solutions*,
https://doi.org/10.1007/978-3-032-21940-4_5

various models developed in response to the AI agent requirements discussed in Chap. 4.

Keywords Software wave · AI prediction and optimization · Machine intelligence · Instruction models · Thinking models · Deep learning model · AI goal achievement matrix model · Discount rate · Net present value · Habitat sustainability index model · Carbon sequestration model · Climate impact assessment model · Local participation model · Early warnings for all · Wetland connectivity · Eco-cultural tourism · Song Dam Lake · Multi-AI agent · Connectivity scoring method

5.1 The Role of Models: Instruction Models Versus Thinking Models

AI optimization refers to the process of enhancing artificial intelligence (AI) models and algorithms to improve their performance, efficiency, and effectiveness. This involves refining AI's 'rules' to make it more accurate, faster and resource-efficient. Optimization can focus on operational efficiency (making AI run smoother) or model effectiveness (making AI better at its task).

Table 5.1 shows the main features of instruction models and thinking models. Thinking models are characterized by problem solving, and multi-step agentic reasoning that is more capable of the most complex reasoning.

5.2 Deep Learning Model

Deep learning is one of the problem-solving models used in city climate planning. It is a thinking model. The diagram in Fig. 5.1 proposes a deep-learning-based prediction and optimization model to optimize the suitability of land for different uses. It is a creation of the AI agent-based ecological modeling process.

In the use of the model, the upper part indicates the deep learning workflow, and the lower part indicates the ecological modeling process. The arrows at the center show how to transfer the relevant values of each step in the ecological modeling process into the deep learning modeling process.

Table 5.1 Comparison of AI instruction models versus thinking models

Period	Characteristics	Project and policy	Approach	Key facts
Stage 1: 2003–2007	**Start of u-City**: • ICT-driven • Sensor-based service • Operational reformation	• Songdo & Busan u-city (2003)	• Bottom-up approach • Development led by municipalities and urban developers	• Success cases in urban services innovations • City as a product approach, leading to slump of u-city
Stage 2: 2008–2010	**Downturn of smart city (smart city winter)**: • ICT-driven • Institutionalization	• Hwaseong-Dongtan u-city (2008) • Special Act on Smart City (2008) • First national smart city plan (2009)	• Top-down approach • Institutionalization led by national government	• National policy framework • Absence of smart city in urban development • Smart city plans, but lack of action
Stage 3: 2011–2016	**Beginning of data-based innovations**: • Data driven • Pillars of smart city innovation	• Smart Seoul 2015 (2011) • Open data and big data for urban innovations (2012) • Second national smart city plan (2013)	• Middle-up-down approach • Innovations led by municipalities • Institutions developed by national government	• Urban innovations not under the name of smart city • Beginning of disruptive innovations
Stage 4: 2017–present	**Rebooting of smart city**: • Platform driven • City transformation	• Smart City Special Committee (2017) • New smart city strategy (2018) • National pilot smart city (2018)	• Semi-top-down approach • Smart city strategy redesigned by national government • Municipality-level master planning and pilot projects	• Smart cities for the fourth industrial revolution • New smart city concept and pilot project

Source Sin (2025)

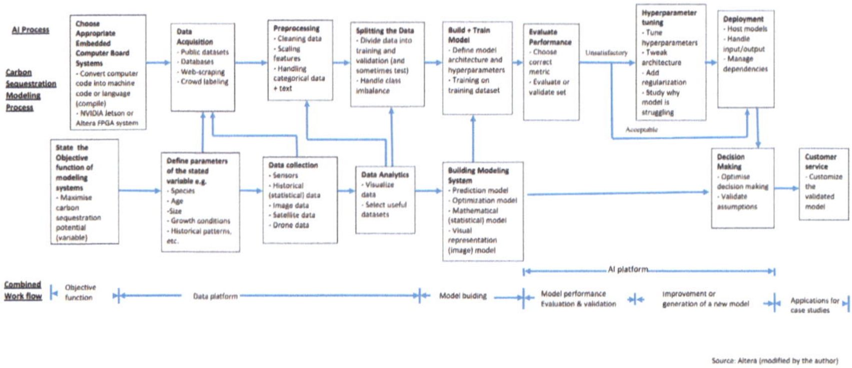

Fig. 5.1 Application of AI deep learning workflow in ecological modeling process

5.3 AI Goal Achievement Matrix Model Using Sustainability Indicators

The sustainability indicator approach to optimization involves using specific metrics to guide decision-making towards more sustainable outcomes. This approach integrates environmental, social and economic considerations into the optimization process, ensuring that solutions are not only efficient but also contribute to long-term well-being and resource conservation.

5.3.1 AI Sustainability Goal Achievement Matrix Models as Project Plan Evaluation Tools

Basics

Continued effort is needed to quantify the intangible and long-term costs and benefits of a particular project and to estimate meaningful discount rates and opportunity costs for sustainability (Canter 1977). The AI prediction and optimization function can help to solve the problems facing traditional cost-benefit analysis.

An AI sustainability cost-benefit analysis (CBA) for a proposed project plan identifies and quantifies the economic, social, and environmental costs and benefits (direct and indirect) against sustainable development goals to make informed, efficient, and sustainable decisions. Key steps include (1) identifying diverse costs and benefits (monetizing them where possible), (2) applying a discount rate for future impacts, and (3) assessing risks to determine if the social benefits outweigh the costs. While useful, traditional CBA can underestimate costs, especially for long-term, cumulative impacts of the proposed projects on climate change for truly transformative change.

Framework

A structured approach to evaluate the climate impacts of projects by comparing their associated costs and benefits is as follows:

(1) Goal setting: Select projects or policies that cause the minimum climate impacts
(2) Identification of costs and benefits: List all potential impacts, both positive and negative, that a project might have, such as:
    Costs: Abatement costs, infrastructure investment, policy implementation expense
    Benefits: Reduced climate damages (from flooding, droughts), improved public health, increased biodiversity, and creation of green jobs.
(3) Monetization of impacts: Assign a monetary value to these costs and benefits to compare them directly.
(4) Application of a discount rate: Convert future benefits and costs to their present-day value to account for the time value of money and evolving societal well-being.
(5) Assessment of net present value (NPV): If the total value of benefits (discounted to the present) exceeds the total value of costs (discounted to the present), the project or policy is considered cost-beneficial.

Goal achievement models in the context of extended cost-benefit can indeed serve as project plan evaluation models, particularly in contexts where the primary objective is to assess the extent to which a plan has successfully met its intended sustainability outcomes. These models focus on measuring the degree to which specific sustainability goals, established during the planning phase, have been attained.

Goal achievement models offer a valuable framework for evaluating plans by focusing on the extent to which intended outcomes have been realized. By clearly defining goals, measuring progress, and analyzing results, these models can provide valuable insights for improving future planning and implementation of projects for climate solutions.

5.3.2 AI Goal Achievement Matrix Model for Carbon Neutrality and Sustainability

A model is proposed for SDGs, and emphasizes evaluating the extent to which sustainable development objectives have been achieved by the proposed project. This model intends to demonstrate how AI can drive sustainable change.

The goal achievement matrix model explores how we can build cities that thrive in the face of climate change, rapid urbanization and much more. This model brings together an interdisciplinary team.

We need to discover what the future of urban sustainability looks like when innovation meets climate impact.

The proposed predictive methodologies involve the application of AI technology. The major result of these methods is an aggregate index of goal achievement for each of the project plan alternatives, so that they can be compared on a common basis (Canter 1977).

The conventional goal achievement matrix models can be extended to show the sustainability cost and benefit of a project to leave a positive carbon footprint. A combined and mixed scanning approach of backcasting and forecasting is used to build up this particular model. This technique can be illustrated by the following sample problem (Table 5.2) in terms of sustainability cost and benefit. Table 5.3 can be utilized to compile a complete and accurate basic data to fill out the AI goal achievement matrix.

Mitigation measures for sustainable development goals

- Targets:
  - 30% reduction in CO_2 emissions in 2030
  - Transitioning to renewable energy sources
  - Improving energy efficiency
  - Reducing waste
- Indicators:
  - Indicator 13.1: Tons of CO_2 emissions reduced within 5 years (2030)
  - Indicator 13.2: kW of increase in renewable energy use
  - Indicator 13.3: Tons of waste reduced
  - Indicator 13.4: kW of energy consumption reduced
  - Indicator 13.5: No. of AI education initiatives
  - Indicator 13.6: No. of integrated and holistic approaches to localized production of energy, water and food
  - Indicator 13.7: No. of green roofs
  - Indicator 13.8: No. of projects integrating small-scale agriculture into urban communities
  - Indicator 13.9: No. of protected natural wetland systems
  - Indicator 13.10: No. of projects restoring, enhancing, and creating wetlands to provide for additional carbon sinks
  - Indicator 13.11: No. of climate policy measures
  - Indicator 13.12: No. of low-carbon buildings
  - Indicator 13.13: No. of climate stable practices
  - Indicator 13.14: No. of transit-oriented developments
  - Indicator 13.15: No. of cycle-city projects
  - Indicator 13.17: Tons of CO_2 emissions reduced through climate responsive urban form.

Adaptation Measures

- Targets:
  - To reduce climate disasters

Table 5.2 AI goal achievement matrix model for project plan evaluation using climate sustainability indicators

Alternative	SDG 13 (Climate Change): take urgent action to combat climate change and its impacts																	
	Mitigation measures adaptation measures																	
	Ratings	Climate sustainability indicators																
		13.1	13.2	13.3	13.4	13.5	13.6	13.7	13.8	13.9	13.10	13.11	13.12	13.13	13.14	13.15	13.16	13.17
Plan Alt. A	No. of plus																	
	No. of minus																	
	Total no. of ratings																	
Plan Alt. B	Algebraic sum of ratings																	
	Average ratings																	
	No. of plus																	
	No. of minus																	
	Total no. of ratings																	

Note 1: Algebraic sum of ratings:
Plus ratings for contributing to goal achievement: +1, +2, +3
Minus ratings for hindering goal achievement: −1, −2, −3
Note 2: Targets and 17 indicators for climate mitigation and adaptation measures

Table 5.3 Data sources and implications for AI learning data: an illustration

Data source	Relevant indicator data	Implications for sustainability
AI cloud access (server)		
ResearchGate access		
Literature review		
Field survey (monitoring)		
Expert knowledge		
Local knowledge		
Review of case studies		

  - To increase resiliency

- Indicators:
  - Indicator 13.18: No. of designated flood risk zones
  - Indicator 13.19: No. of historic settlements and individual assets and archaeology
  - Indicator 13.20: No. of protected areas affected by flooding and coastal erosion and inundation.

It is essential to ensure that the overall indicators of plan alternatives should be fed through other delivery mechanisms such as the local development framework. It will also be essential to monitor the sustainability outcomes that can be attributed to the indicators. This will help to determine at as early a stage as possible whether the plan is being delivered in accordance with the indicators, and if not, where corrective action needs to be taken.

An organization might map its efforts to reduce greenhouse gas emissions (SDG 13) to specific activities like transitioning to renewable energy sources, improving energy efficiency, and reducing waste. They would then select indicators like tons of CO_2 emissions, energy consumption and waste generation to track progress. Targets might include a 20% reduction in emissions within five years, or a 10% increase in renewable energy use within three years.

By using a sustainability goal achievement matrix, organizations can effectively integrate sustainability into their core operations, track their progress and contribute to a more sustainable future.

5.4 AI Habitat Sustainability Index Model

Habitat suitability analysis (HSA) is a method for predicting where a particular species of wildlife is most likely to occur and thrive, based on the environmental factors it selects for or avoids.

This modeling technique assesses landscape quality and predicts a species' potential spatial distribution, helping to prioritize conservation and restoration efforts.

5.4.1 Habitat Suitability Analysis: Overview

A habitat suitability analysis uses a statistical modeling approach to correlate species observation data with environmental variables. The process generally follows these steps:

(1) Select key wildlife target species (e.g. stork).
(2) Identify habitat requirements: Researchers define the critical environmental and ecological factors that influence the target species' survival (eating, hiding, feeding, daytime activities, sleeping) and reproduction.
(3) Collect and prepare data: Data on the species' occurrences and the relevant environmental factors (known as 'predictor variables') are gathered.
(4) Create a common scale: The different criteria are standardized to a common suitability scale, often ranging from 0 (unsuitable) to 1 (highly suitable).
(5) Weight and combine criteria: The different variables are given weighted values based on their relative importance to the species and combined to create a habitat suitability index (HSI) map.
(6) Identify suitable areas: The resulting HSI map highlights areas with varying degrees of habitat suitability, from unsuitable to highly suitable. The mapping can be done by a multi-AI agent.

5.4.2 HSA Key Data and Variables

Habitat suitability analysis integrates a wide range of data, often using Geographic Information Systems (GIS) to analyze layers of spatial information.

- **Species occurrence data**: Records of where the species has been observed. This historical learning data can be used to predict a species distribution.
- **Topographic factors**: Elevation, slope and aspect.
- **Land cover and vegetation**: Types of vegetation, land use and the density of plant cover.
- **Hydrological features**: Proximity to water sources like rivers and lakes.
- **Climatic variables**: Temperature and precipitation patterns.
- **Anthropogenic factors**: Human population density, distance to roads, and disturbances.

5.4.3 Applications of HSA for Goal Achievement Matrix

This section proposes a goal achievement matrix model for project plan evaluation using habitat suitability indicators. This model uses a multi-AI indicator decision-making technique to assign values to different habitat suitability indicators. The steps are as follows:

(1) Selection of SDG for sustainable habitat management: SDG 15, 'Life on Land'. This goal focuses on protecting, restoring and promoting the sustainable use of terrestrial ecosystems. This includes sustainability managing forests, combatting deforestation, halting and reversing land degradation and protecting biodiversity.
(2) The targets:

   a. Conserve and restore terrestrial and freshwater ecosystems
   b. End deforestation and restore degraded forests
   c. End desertification and degradation
   d. Protect biodiversity

(3) The key habitat sustainability indicators:

   a. Indicator 15.1: Forest area as a proportion of the total land area
   b. Indicator 15.2: Proportion of important sites for terrestrial and fresh water biodiversity that are covered by protected areas, by ecosystem type.
   c. Indicator 15.3: No. of mitigation proposals
   c. Indicator 15.4: No. of ecological corridors (flyways)
   d. Indicator 15.5: Proportion of feeding, hiding, mating, sleeping and playing activity areas
   e. Indicator 15.6: No. of measures to avoid impacts
   f. Indicator 15.7: No. of alternative sites for conservation
   g. Indicator 15.8: No. of animal and plant population estimates.

A Hypothetical Example Model

This technique is illustrated by the following sample problem. The steps involved in the use of matrix in Table 5.4 are as follows:

(1) Read different habitat suitability area maps in the plan alternatives at Y axis
(2) Read key habitat suitability indicators to the left
(3) Indicate ratings of plan alternatives against habitat suitability indicators.

5.5 Carbon Emissions and Sequestration Models

AI models such as machine learning (ML) are used for carbon emissions reduction and sequestration.

Table 5.4 AI goal achievement matrix model for land use plan generation and evaluation using habitat suitability indicators

Alternative	Ratings	SDG 15							
		Key habitat suitability indicators							
		15.1	15.2	15.3	15.4	15.5	15.6	15.7	15.8
Plan Alt. A (No action alternative)	No. of plus								
	No. of minus								
	Total No. of ratings								
	Algebraic sum of ratings								
	Average ratings								
Plan Alt. B	No. of plus								
	No. of minus								
	Total No. of ratings								
	Algebraic sum of ratings								
	Average ratings								
Plan Alt. C	No. of plus								
	No. of minus								
	Total No. of ratings								
	Algebraic sum of ratings								
	Average ratings								

Note Algebraic sum of ratings

- Plus ratings for contributing to goal achievement: +1, +2, +3
- Minus ratings for hindering goal achievement: −1, −2, −3

5.5.1 AI Models for Optimizing Solutions to Reduce CO_2 Emissions

The artificial intelligence models for reducing CO_2 emissions are required to:

- Increase the use of AI models for optimizing solutions to reduce CO_2 emissions.
- Train and use AI models to generate CO_2 emissions.
- Quantify positive and negative climate impacts of AI models.

Nowadays, there is an increasing use of digital technologies and AI such as ML models that leverage data to optimize the performances of systems in almost all activity sectors, including ML models for optimizing solutions related to CO_2 capture from the atmosphere or CO_2 emissions reduction from human activities.

However, the use of AI models is also leading to increased energy consumption and CO_2 emission issues, as discussed in Chap. 4.

5.5.2 A Method for Assessing Carbon Sequestration

Machine learning (ML) is a powerful computational tool that is transforming the way we manage large carbon sinks. AI refers to computer systems designed to perform tasks that typically require human intelligence, while ML is a subset of AI where algorithms learn from data to improve their performance without explicit programming. When applied to managing carbon sinks, these technologies offer unprecedented capabilities for analyzing vast amounts of data, making predictions and providing insights to inform decision-making.

Utilizing AI and ML for managing large carbon sinks promises to revolutionize climate action. By optimizing carbon sequestration and enhancing monitoring techniques, it enables more effective stewardship of vital ecosystems. This innovation enhances our capacity to mitigate climate change, ensuring sustainable management of critical carbon reservoirs for a healthier planet.

Carbon sequestration algorithms, predictive models and optimizing model include the following:

(1) **ML algorithms**: ML algorithms, such as deep learning, random forests, and support vector machines, are used to analyze data from various sources, including:
   - **Remote sensing data**: Satellite imagery, LiDAR data, and hyperspectral images provide valuable information about forest cover, wetland cover, vegetation health, and carbon content.
   - **Ground-based measurements**: Field surveys, soil samples, and sensor networks provide detailed information about specific locations.
   - **Environmental data**: Climate data, soil maps, and topographic information provide contextual data for understanding carbon sink dynamics.

(2) **AI-powered predictive models**: By learning from historical data, AI can develop predictive models to Forecast Carbon Sequestration Potential (refer to Table 4.4).

Carbon Sequestration Optimizing Model: Hypothetical Example (Linear Computer Programming)

Step 1: The objective function

(1) To maximize the total additional carbon sequestration value

Step 2: The constraints

(2) To maximize the area of land to be restored (a land constraint).

The land constraint means that the area of plant communities restored must not use more than available supply of land.

As an example, it provides that:

$$\frac{1}{10}\mathbf{A} + \frac{1}{5}\mathbf{B} \quad \text{must not exceed 50 acres}$$

Land area required for plant A species	Number of plant A species	Land area required for plant B species	Number of plant B species

(3) To minimize the cost of wetland restoration (the budgetary constraint)

The budgetary constraint means that the total cost of plant communities restored must not exceed the available finance.

As an example, it provides that:

2000 **A** + 6000 **B** must not exceed $1,200,000

(4) To combine the two constraints above, alternatively, these may be expressed as:

$$\frac{1}{10}\mathbf{A} + \frac{1}{5}\mathbf{B} \leq 50 \text{ acres}$$

$$2\,000\,\mathbf{A} + 6\,000\,\mathbf{B} \leq \$1{,}200{,}000$$

$ 2000 : cost of restoration of plant A species	$ 6000 : cost of restoration of plant B species

Step 3: The optimum policy

The optimum policy is represented by the point in the feasible area that provides the greatest possible carbon sequestration value (for the expression (equation) 190**A** + 470 **B**, which satisfies the land and budget constraints).

The essence of a linear programming computation with AI is the search for this point. A computer program determines this point by algebraic methods, perhaps with more and different variables.

5.6 AI Applications in Climate Impact Assessment

5.6.1 *Downscaled Climate Impact Assessment Model: Overview*

Climate impact assessment (CIA) should be an integral part of the city climate planning at the initial stages of project planning. Future trends indicate a focus on the city climate impact assessment model and greater local participation. Detailed requirements will be necessary in the future in response to the degree that climate impact assessment becomes a part of project planning documentation.

5.6.2 *Key Stages of a Climate Impact Assessment*

(1) **Screening**: This initial stage determines if a full climate impact assessment is necessary based on the project's potential for significant climate impacts.
(2) **Scoping**: If a full assessment is required, scoping defines the boundaries of the assessment, identifies key issues and sets the basis for analysis.
(3) **Impact assessment and evaluation**: This stage involves identifying, predicting, and evaluating the potential climate, environmental and social impacts of the project, including both positive and negative effects.
(4) **Mitigation and management**: This involves developing strategies to avoid, minimize, or offset negative impacts, and enhancing positive impacts. This may include changes to the project design, implementation of mitigation measures, and development of a Climate Management Plan (CMP).
(5) **Monitoring**: This stage involves tracking the implementation of mitigation measures and assessing the effectiveness of the CMP, ensuring ongoing management of climate impacts.
(6) **Public participation**: Throughout the process, public consultation and participation are crucial, ensuring that the concerns of affected communities are addressed.
(7) **CIA report**: A comprehensive report summarizing the findings of the assessment, including the identified impacts, mitigation measures, and monitoring plans, is prepared for review and decision-making.

5.6.3 *Need for Applications of AI in CIA*

The sustainability indicator system described in Chap. 4 can be used in the downscaled climate impact assessment at the community, city and regional level. AI is becoming increasingly valuable in climate impact assessments by improving data analysis, prediction and optimization across various sectors. It helps in faster risk assessments, better resource management and the development of more effective mitigation and adaptation strategies.

5.6.4 Addressing the Climate Impact of AI

While AI offers significant benefits for climate impact assessment, it's crucial to acknowledge its own climate impact, particularly in terms of energy consumption and water resource usage in data centers.

Efforts are underway to measure and mitigate the carbon footprint of AI, including optimizing data center efficiency, decentralization of datacenters, and promoting the use of renewable energy. It is important to address the climate impact of AI itself and ensure its responsible development and deployment in the CIA process.

5.6.5 Methods of AI Technology for CIA

Downscaled climate impact assessment (DCIA) can embrace the AI age, while still retaining some traditional methods. Planners can develop data collection and analysis, and impact prediction, assessment and evaluation models as testing models, later developing complex models using an AI palette of software that includes AI algorithms to predict, assess and optimize the impacts and visualization. The use of AI means as a way to prepare climate impact assessment statements is progressive move in the field of climate impact assessment. A wide range of AI tools can be used in the whole process of CIA—including NDVIA Jetson, Deep Learning workflow, Google Cloud, ASUS Intelligent Edge Computing and AI Solutions, ADLINK Edge Computing Platforms, NUBISON AIOT and Ceph Multicloud K8s, to name just a few (Table 5.5). The AI tools allow us to expand our thinking in creating a CIA project. AI software can be categorized into various types, including generative AI (text, images, code), analytical tools, automation tools and city-specific tools. Popular examples include ChatGPT, DALL-E, IBM Watson, Zapier and PathAI. Specific AI software categories include chatbots, AI search tools, AI assistants and image generators. The following table presents use of AI for the CIA process to utilize data of potential project activities.

5.6.6 AI Applications in the CIA Process for City Climate Planning

In order to facilitate better assessment of the impacts of proposed projects on climate change, we try to develop and maintain downscaled climate impact assessments,

Table 5.5 Use of AI in the climate impact assessment process

CIA stages	Descriptions	AI applications (AI Software Programs)	Attainable or unattainable
1. Description of the project proposal	• Location • Size • Time period • Method of construction		• Knowledge gaps identified in the AI transformation of CIA
2. Description of the existing environment	• Types of data – Physical-chemical data – Biological data – Cultural data – Socioeconomic data • Data analysis • Remote sensing • Computer mapping approaches • Data storage and retrieval systems	• AI data center • AI data analysis • AI data platform • Data security • IBM Watson (data insights) • Tableau (data visualization) • Microsoft Azure storage, Virtual Machine (VM), Cloud, Services, and HDInsight • Ceph storage (Ceph Multi-cloud K8s) • Amazon Web Services	The deployment of CO_2 sensors might be too sparse to report the existing CO_2 situation in the site.
3. Impact prediction	• The primary and secondary effects of project development and implementation • Identification of environmental modifications that may be significant • Forecasting of the quantity and/or spatial dimensions of changes in environment identified • Estimation of the probability that the impact (environmental and climate change) will occur (time period)	• Urban simulation • 3D visualization • AI land use change models: multiple regression analysis • 1.5 °C climate change modeling • SDA solution • NVIDIA Jetson • Altera FPGA system • NUBISON AIOT • IBM Watson (predictive analytics) • Google Cloud's VertexAI (machine learning models) • Microsoft Copilot Studio (customization of Copilot for specific needs)	Without climate big data, deep learning-based predictive models may not be useful Other sensing technologies such as crowdsourcing-based applications using mobile phones could be used

(continued)

Table 5.5 (continued)

CIA stages	Descriptions	AI applications (AI Software Programs)	Attainable or unattainable
4. Assessment (evaluation)	• Extended sustainability cost–benefit analysis to quantify the intangible and long-term costs and benefits of a particular project and estimate meaningful discount rates and opportunity costs • Land carrying capacity analysis using climate sustainability indicators • Extension of risk analysis • Assessing the cumulative impacts of new land uses • Quantitative and qualitative standard assessment	• Microsoft Copilot Studio (customization of Copilot for specific needs)	
5. Selection of mitigation measures and mitigation-based alternatives: optimization	• Incorporating environmental and climate factors into planning process – Checklist methods – Matrices methods – Network analysis methods • Project alternative evaluation methods – Sustainability cost–benefit analysis – Goal achievement matrix	• Autodesk InfraWorks • AI impact operation management • AI goal achievement matrix using sustainability indicators	It is of great importance to the sustainability of the project if we can optimize mitigation measures over the entire life cycle of the project, including supply chain emission through smarter procurement towards net zero.

(continued)

Table 5.5 (continued)

CIA stages	Descriptions	AI applications (AI Software Programs)	Attainable or unattainable
6. Climate monitoring	• Climate data collection • Climate data processing • Climate data analysis	• Integrated AI system for performance verification for impact assessment • Envirosuite • IBM Watson Environmental Compliance • Google Earth Engine • EarthAI • Autodesk InfraWorks	
7. Public participation	• Incorporating of public participation through the process leading to a decision regarding a proposed action • Forward and feedback techniques	• Perplexity (citizen science) • Client service AI • Personalized service AI • Use of chatbot and AI-aided visual methods for presentations for residents on climate impacts	

based on the latest site observation and modeling data developed in this book. The downscaled data and assessments could be incorporated into the system described in the Framework for City Climate Planning AI, and used to analyze proposed projects and land-use decisions to improve the resiliency of new projects.

It can be claimed that AI climate impact assessment will be the future of the Greenhouse Gas Protocol to promote AI transformation of cities for climate solutions.

5.7 AI Local Participation Model

Citizen science is a form of scientific research in which members of the community participate alongside professional planners and scientists. This can involve various activities at some or all stages in a project cycle, such as data collection programs, analysis, design of the project, construction, operation and maintenance, monitoring and evaluation and collective decision-making. The aim is to engage the community in the whole planning and development process to address real-world problems, generate new knowledge and promote the usage of personalized AI models and tools in the surrounding communities. Table 5.6 introduces many new perspectives on AI climate solutions at the community level. The use of AI could include any or all of the components shown in the following table.

5.8 AI Climate Change Adaptation Agent Model: A New Age for AI Defence Strategy with Early Warning Systems

At COP30, one key issue was climate adaptation. COP30 offered the opportunity for countries to set a new adaptation finance goal. Amid record heat, devastating floods and storms, and increasingly destructive fires, many are calling for a commitment to triple the amount of funds for climate resilience (Climate Action News 2025).

5.8.1 Background of UN Early Warnings for ALL

The UN Early Warnings for All (EW4All) initiative identifies several recent efforts in the climate change adaptation policies and measures to be included in adaptation action plans.

As part of the UN Secretary-General's Acceleration Agenda, the Early Warnings for All initiative is a key contribution to delivering climate justice to those at the frontlines of the climate crisis. It aligns with the priorities of the Paris Agreement and supports key provisions of the Sendai Framework for Disaster Risk Reduction, particularly Target G on availability and accessibility of multi-hazard early warning systems. It also contributes to delivering the targets of the 2030 Agenda for Sustainable Development on poverty, hunger, health, water, clean energy, climate action and sustainable cities.

Guterres (2022) states that:

> Today, one third of the world's people, mainly in least developed countries and small island developing states, are still not covered by early warning systems. This is unacceptable, particularly with climate impacts sure to get even worse. Early warnings and action save lives. To that end, today I announce the United Nations will spearhead new action to ensure every person on Earth is protected by early warning systems within five years.

Early Warnings for All (EW4All) is built on four pillars that are the cornerstones of the initiative and of effective multi-hazard early warning as follows:

- Disaster risk knowledge and management (led by UNDRR)
- Detection, observation, monitoring, analysis, and forecasting (led by WMO)
- Warning dissemination and communication (led by ITU)
- Preparedness and response capabilities (led by IFRC).

Table 5.6 Use of AI in the project cycle stage of local participation

Project cycle stage	Model local participation plan	AI technology applications
Planning and design stage	• Preparation of community maps (a graphic representation of the community or specific aspects of it) to uncover the community's self-perceptions • Understanding of characteristics of local people: voicing their opinions, educational background, etc.	• AI visual representations (e.g. digital twin, metaverse, space intelligent technology, etc.) • AI online information-sharing materials using 5G/6G • Generation of AI text and image information (space information and digital information)
	• Identification of a list of principal problems with the project by the community	• Digital shadow methods
Construction work stage	• Development of the goals and objectives of the project to build consensus as early as possible: leading voice	• AI high-tech architecture
	• Collection of data on local and indigenous knowledge on the site: local-led conversation	• Reporting and dashboard management • Meat and dairy emission dashboard
	• Preparation of master plan to accurately consider and weigh the various physical, ecological and socio-economic indicators to establish climate, vegetation, water, and land carrying capacities	• AI real-time visitor counter • Use of Deepseek, Chatbot and ChatGPT for nature interpretation
Operation and maintenance stage	• Use of local building materials for healing ecology	
	• Daily visitor count	• AI big data analytics • AI prediction models
Monitoring and evaluation stage	• Monitoring of the implementation of the project and evaluation of the effectiveness of its operation for adjustments which can be made to address unforeseen problems or circumstances	• AI optimizing models • AI business 5.0 and finance analysis
Collective decision-making	• Collective decision-making which would come to a consensus on actions taken in response to the impacts of the eco-cultural tourism project, including climate impact analysis • Building of local environmental stewardship and climate watchdog	• Multilingual communication program • Big digital platforms and virtual community

To ensure progress and the continued strategic alignment of activities with implementing bodies, the UN Secretary-General created the Early Warnings for All Advisory Panel, co-chaired by the Executive Heads of WMO and UNDRR, for the duration of the Initiative. The Advisory Panel consists of the Heads of a multitude of UN organizations, the private sector, civil society and UN Member States. Meetings of the Advisory Panel occur biannually, and the Panel prepares an annual progress report to the Secretary-General on the status of the initiative's activities.

The EW4All initiative has been launched at the national level without considering proper AI applications in the early warning system. UNFCCC is not directly engaged in the initiative, as can be seen in the four pillars above. So, we need a downscaled action-oriented AI city initiative at the local or regional level to support the EW4All under the leadership of UNFCCC.

5.8.2 AI Applications for EW4All

An AI early warning system can use artificial intelligence to analyze vast amounts of data to detect patterns and provide advance notice of potential threats, such as natural disasters, disease outbreaks, or safety issues. These combined systems enhance traditional methods by accelerating the detection and prediction process, allowing for faster responses to mitigate negative impacts. They work by integrating diverse data sources and are most effective when used as a tool to support, not replace, human expertise.

Table 5.7 shows an AI multi-hazard approach to handle multiple hazards, as climate change intensifies a range of threats. This table is based on the four pillars of EW4All, and the composition of climate change and disaster risk shown in Fig. 5.2. Downscaled data inadequacies should be overcome in the context of city and region.

Table 5.7 AI-based climate change adaptation: early warning system AI agent model

The five pillars of strategic activities	AI applications
Data streams (detection, observation, monitoring, analysis, forecasting)	AI systems to process and integrate diverse data streams into real time, including meteorological, geological, social media and historical records, etc.
Pattern recognition	Machine learning algorithms to identify complex patterns and anomalies
Disaster risk knowledge and management	AI predictive models to forecast the potential impact and severity of an event
Warning dissemination and communication	AI systems helping to streamline the communication of warning to vulnerable populations, aiming for timely and efficient alerts.
Preparedness and responses	AI systems helping to improve disaster preparedness and responses

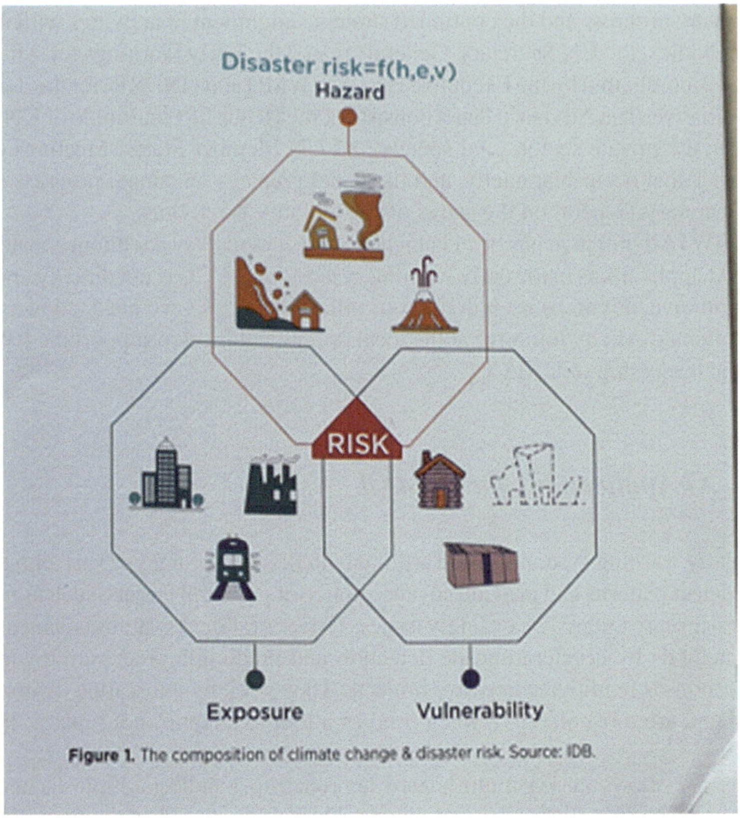

Fig. 5.2 The composition of climate change and disaster risk. *Source* IDB (2018)

It is important to recognize that we need to be integrated system thinkers who can effectively communicate, coordinate, and collaborate with a range of experts and stakeholders in order to effectively implement the array of AI applications listed in Table 5.7.

5.8.3 Recommendations for Participation in the Movement

It is hoped that the International Advisory Group for AI City Networking Initiative will be part of the UN movement and engaged in the AI proposal in partnership with the UN EW4All Initiative and UNFCCC Climate Change AI Forum in partner cities.

5.9 Integrated Solutions for Wetland Connectivity in AI-Based Eco-Tourism

AI and wetland science meet at the Song Dam Lake, Da Nang City, Vietnam. Wetlands are critical ecosystems providing ecological, cultural, and economic benefits. However, fragmented management and climate change have degraded their connectivity and resilience. This paper proposes an integrated AI-based framework to enhance wetland connectivity and support eco-cultural tourism development, using Song Dam Lake in Da Nang City, Vietnam as a case study area. The first part describes connectivity theory, the best management practices of wetlands and the role of digital technologies in connecting fragmented wetlands. The second part proposes a new innovative methodology which integrates IoT-based monitoring, AI habitat modeling, carbon assessment and community engagement through multi-agent systems. The study highlights how digital connectivity, when coupled with ecological planning, can promote sustainable wetland management and climate-smart tourism. Findings demonstrate the potential of AI in linking ecological integrity with socio-economic development for wetland-based communities.

5.9.1 Introduction

Background

Wetlands are often described as the 'green spines' of the planet, sustaining biodiversity, regulating hydrological cycles and supporting human livelihoods. However, traditional case-by-case approaches to wetland management have frequently led to fragmented decision-making and the gradual degradation of wetland ecosystems. The lack of integration among ecological, hydrological and socio-economic datasets weakens systemic understanding, limiting efforts to maintain connectivity and resilience across wetland landscapes.

In the era of the 4th industrial revolution, digital technologies such as Artificial Intelligence (AI), the Internet of Things (IoT) and cloud computing are transforming environmental monitoring and management. When combined with ecological principles, these technologies enable dynamic, data-driven and predictive approaches to conservation. The integration of digital connectivity with ecological connectivity allows wetlands to be managed as interconnected systems rather than isolated sites, strengthening both ecosystem resilience and socio-economic sustainability.

Emerging studies have demonstrated the growing potential of AI and IoT in wetland applications. For instance, Lin et al. (2019) applied deep convolutional neural networks to classify wetland vegetation morphology with high precision, while Nundloll et al. (2019) developed IoT-based infrastructures for real-time monitoring of water quality and soil moisture in remote wetlands. These innovations redefine how wetland ecosystems are observed, modeled and managed—particularly in regions exposed to climate change, salinization and rapid urbanization.

In Vietnam, the Song Dam Lake ecosystem in Da Nang City exemplifies both the ecological importance and vulnerability of coastal wetlands. Covering approximately 200 hectares, it serves as a crucial freshwater source, habitat for migratory birds and a center of cultural identity for local communities. Yet, increasing salinity, vegetation loss and uncoordinated land-use practices have weakened its ecological integrity and connectivity.

Objectives

This study aims to develop and conceptualize an AI-based integrated framework and methodology for sustainable wetland management at Song Dam Lake. The proposed framework seeks to:

- Integrate ecological, hydrological and digital data layers to enhance wetland connectivity and resilience
- Apply AI and IoT technologies for real-time environmental monitoring, habitat suitability modeling and carbon assessment
- Promote community engagement and eco-cultural tourism through AI-assisted participatory platforms
- Demonstrate a replicable model of smart, climate-resilient, and community-driven wetland management for developing regions.

By combining scientific assessment, digital innovation and local participation, the framework provides a holistic foundation for AI-assisted wetland restoration and eco-cultural tourism development, ensuring that the Song Dam Lake system remains both ecologically vibrant and socio-culturally meaningful.

5.9.2 Issues Related to Wetland Planning and Management

Water as a Precious Resource

Water is the lifeblood of wetland systems, forming the foundation of ecological structure and function. Effective wetland management requires a comprehensive understanding of hydrological processes across spatial and temporal scales. Yet, in many developing regions, water management is often sectoral and fragmented, neglecting the interconnectedness of wetlands, rivers and groundwater systems.

From a systems perspective, wetlands can be perceived as multi-layered ecological entities where hydrology, vegetation and human interactions operate horizontally and vertically. Integrating these layers through smart water grids can promote adaptive and sustainable management.

Horizontal and Vertical Layering of Wetlands

Wetland systems are characterized by two main dimensions of layering:

(1) Horizontal layering reflects the spatial arrangement and hydrological connectivity among wetlands—from isolated wetlands to connected wetland networks.

Geographically isolated wetlands lack surface-water connections to downstream rivers but still provide crucial ecosystem services such as groundwater recharge and biodiversity support (Tiner 2003).

Connected wetlands, unlike isolated wetlands, are wetlands that exhibit clear hydrological and ecological connections with rivers, lakes or other wetland systems. This connectivity plays a critical role in maintaining water flow, nutrient cycling and migration corridors for aquatic flora and fauna. In the example of Barn Elms, London (UK), a green–blue grid system has been implemented to reconnect people with nature while enhancing resilience to flooding and climate change.

(2) Vertical layering refers to the ecological and functional stratification within a wetland including soil moisture gradients, vegetation structure and biogeochemical processes. Wetlands have multilayers of structure and function based on tradition and organic fabric ecosystems

The diagram in Fig. 5.3 is divided into two main sections:

(1) *Left side*: Three main policy objectives in Integrated Water Resources Management (IWRM)

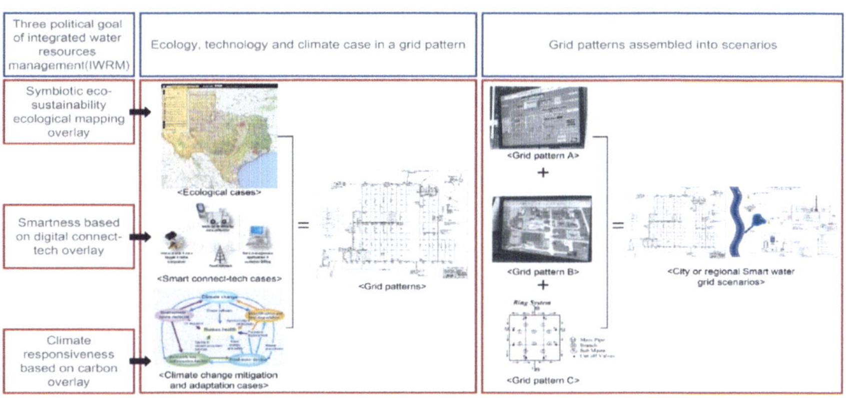

Fig. 5.3 Alternate smart water grid development scenario model: the nature of technology should be symbiotic not unidirectional, and low-carbon and less vulnerable technology applications (Kim 2017)

a. Symbiotic eco-sustainability ecological mapping overlay: Emphasizes the harmony between human development and ecological conservation.
b. Smartness based on digital connect-tech overlay: Application of IoT, AI, and big data to monitor and manage ecosystems in real time.
c. Climate responsiveness based on carbon overlay: Integration of climate mitigation and adaptation solutions, such as carbon sequestration, flood planning, and emissions management.

These three layers are overlaid to create grid patterns.

(2) *Right side:* Combination of grid patterns into planning scenarios

a. Grid patterns (A, B, C) represent different configurations or management models (e.g., ecological grids, technological grids, carbon grids).
b. Combining multiple grid patterns forms city—or regional-level smart water grid scenarios.

In other words, modern water resources planning does not rely solely on ecology, technology or climate individually, but requires the integration of all three factors into a 'smart grid' structure. This approach enables the development of cities and regions that are resilient, sustainable and capable of effective water management.

5.9.3 *Theoretical Framework*

Innovative Integrated Water Resource Management

Wetlands can generally be described as transitional areas between land and deep-water habitats. Table 5.8 shows attempts for reorienting water resource planning and management towards integrated water resource management (IWRM) in combination with traditional water management and cutting-edge ICT such as IoT and AI technology.

5.9.4 *Exemplar Model of Wetland Planning and Management*

Thinking/Philosophy: Reductionist Approach Versus Holistic Approach

A new approach is needed to rethink how to manage wetlands throughout a watershed.

Table 5.9 shows different approaches at the different level of connectivity.

Connectivity should be considered in the conservation context:

> Connectivity conservation is defined using biodiversity conservation criteria, but also includes social and institutional dimensions. Connectivity conservation describes actions

Table 5.8 Exemplar of SMART innovation challenges (Kim 2017)

	Monolithic water model	Conventional smart water grid model	Innovative smart water grid model
City applications	• Wide application exist in techno-cities	• Increasing applications exist in smart cities	• New applications are encouraged in climate-smart cities
Thoughts and driving forces	• Individual technique-oriented thinking at the site level • Technology individualism • The 3rd industrial revolution	• System thinking at the area wide level connectivity • Techno-optimism • Post–2015 SDGs	• Multiple water resources management thinking that the urban and regional watershed level • Smart climate urbanism • Holistic collectivism • Paris climate agreement • The 4th industrial revolution
Goals	• To promote single technology-oriented water business and industry	• To provide more efficient and smart connected systems between surface water, drinking water, hot water, processed water, treated water, resource water and wastewater plants	• To build expanded connections between water grids, eco-grids and climate grids for more sustainable, resilient, and circular water production, consumption and resource water management system
Network structures: framework	• Disconnected, spot or sectoral network structures rather than fully integrated and coordinated network of water systems • Uni-directional framework	• Limited to optimize network structures between water producers and consumers • Bi-directional framework	• Expanded network structures to combine ecology, water and ICT infrastructures • Multi-directional framework
Technology applications: components	• Use of a computing platform technique as set of processing machine in a single dedicated computer room working in silo, single 'Virtual' computation program	• Optimization techniques for processing data in integrated 'Virtual' computation program	• Techniques to maximize potential of ecology-water-energy-climate connections towards greatest assets with more of advanced technology such as IoT, big data, satellite imagery, computing AI, etc.
Potential markets, job opportunities, cost of time	• Isolated or fragmented infrastructure developments are case of single technology market and a job opportunity • Short-term payback period with environmental and climate externalities	• Digital job opportunities are provided by three digital water grids in the area of economic activation with new technology markets • Medium-term payback with less environmental and climate externalities	• More job opportunities could be increasing considerably by integrated water grid in the area of digitalized climate economy • Long-term payback with high level of integration of technology, environment, energy and climate

Table 5.9 Reduction approach versus holistic approach to connectivity enhancement

Planning and design level	Reductionist approach	Holistic approach
Strategic level (or multiple wetlands in total catchment level)	No	Yes At the 1st level of connectivity (Refer to Fig. 5.4)
Site-specific level (or individual wetland)	Yes At the 2nd level of connectivity (Refer to Fig. 5.4)	Yes At the 2nd level of connectivity (Refer to Fig. 5.4)

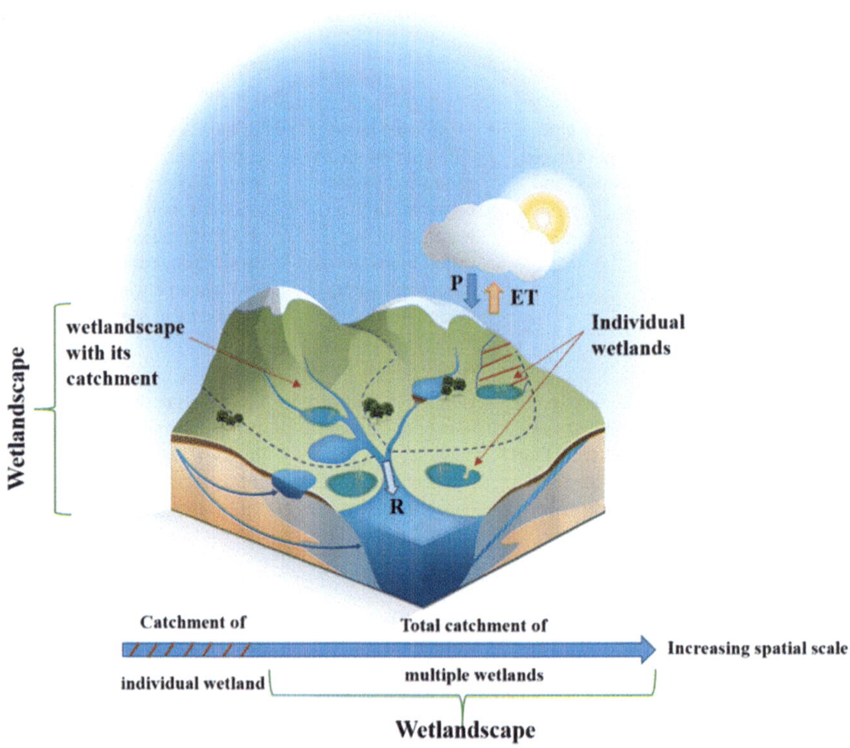

Fig. 5.4 A wetlandscape: the associated large-scale water fluxes (precipitation (P), evapotransportation (ET), runoff (R) and groundwater (GW) flows) are represented by arrows in the landscape

taken to conserve landscape connectivity, habitat connectivity, ecological connectivity or evolutionary process connectivity for natural and semi-natural lands that interconnect and embed established protected areas. (Graeme et al. 2010).

Assessing connectivity can be done at different levels:

- Geographical connectivity: Eurasia + Northeast Asia + Inter-Korea
- Topographical connectivity: Mountain + Hill + Plain, etc.

- Hydrological connectivity: Surface water + Ground water + Precipitation
- Habitat connectivity: Forest + Grassland + Wetland + Open water, etc.
- Wetland habitat connectivity: Inland freshwater marsh + Riparian wetland + Mountain meadow + Peatland + Deepwater swamp + Tidal salt marsh + Tidal freshwater marsh.
- Program connectivity: Forest REDD + Program + Ramsar program, etc.
- Ground connectivity: Breeding grounds + Resting grounds + Feeding grounds + Activity grounds
- Species source connectivity: Species sources + Restored areas, etc.

Wetland Types and Their Connectivity

A wetlandscape consists of multiple wetlands. Figure 2 presents a conceptual sketch of a wetland-scape.

The illustration (Fig. 5.4) shows a 'wetlandscape', a wetland landscape in which multiple small wetlands (individual wetlands) are interconnected to form a system. Each of these individual wetlands is located within a common catchment, and together they constitute the total catchment of the entire wetland landscape.

Connected Wetland Model with ICT/IoT and Blockchain

The 4th industrial revolution and digital transformation including Distributed Ledger Technology (DLT) provide many opportunities for enhancing wetland connectivity with digital technologies. We need a standard paradigm shift. Schwab (2016) states that:

> We are at the beginning of a revolution that is fundamentally changing the way we live, work and relate to one another. In its scale, scope and complexity, what I consider to be the fourth industrial revolution is unlike anything humankind has experienced before." (Schwab, 2016).

Digital technologies and AI integration can be summarized as follows:

- Creation of new value through connecting many different kinds of wetlands
- Open Access Platform

    Microsoft Ajure (IoT program)
    IBM hyperledger (Blockchain program)

- Unit technologies

    Cloud Computing
    AI
    IoT
    Blockchain
    Sensors

Table 5.10 Illustrative applications of ICT and innovation tech in selected wetland management frameworks

Wetland management activities	Smart technologies	
Wetland inventory (survey) for preparation of wetland inventory maps	Drone Sensors Meters	Blockchain CCTV camera
Regular environmental and ecological monitoring	Drone Sensors	Blockchain CCTV camera
Data modeling (algorithm development) Early warning system (prediction and evaluation or ecological changes)	Big data and analytics IoT	AI Cloud computing
Management	Robots	Smart controllers
Education and local citizens' participation in wetland management	Cloud computing	Open access platform

– System Technologies

e.g. Smart controllers which adjust watering cycles using data from ground and weather sensors.

Potential use of smart-tech in wetland management is shown in Table 5.10.

5.9.5 *International Models*

Canadian River Watershed, New Mexico, USA

The Canadian river watershed project in northeastern New Mexico represents a pioneering application of the landscape-level wetland functional assessment (LLWFA) approach in North America. Conducted by Saint Mary's University of Minnesota, Geospatial Services, in collaboration with New Mexico Environment Department (NMED) and US Fish and Wildlife Service (USFWS), the study developed a multi-tiered wetland mapping and assessment framework that integrated GIS, remote sensing, and functional ecology (Saint Mary's University of Minnesota 2013).

Following the Environmental Protection Agency (EPA) (2006) wetland assessment framework, the project mapped and analyzed wetlands across 9,100 square miles within the Upper Canadian, Upper Rio Grande, and Upper Pecos River basins. It extended the traditional National Wetland Inventory (NWI) by integrating hydrogeomorphic (HGM) and functional metrics, linking physical wetland features to ecosystem services. Three assessment levels were applied:

- Level 1: Landscape Assessment: Use GIS and remote sensing to gain a landscape view of watershed and wetland condition. Typical assessment indicators include wetland coverage (NWI), land use and land cover

- Level 2: Rapid Wetland Assessment: Evaluate the general condition of individual wetlands using relatively simple field indicators. Assessment is often based on the characterization of stressors known to limit wetland functions e.g. road crossings, tile drainage, ditching.
- Level 3: Intensive Site Assessment: Produce quantitative data with known certainty of wetland condition within an assessment area, use to refine rapid wetland assessment methods and diagnose the causes of wetland degradation. Assessment is typically accomplished using indices of biological integrity or hydrogeomorphic function.

Using the LLWW (Landscape Position-Landform-Waterbody Type-Water Flow Path) classification systems with HGM indicators, the project generated a functional correlation matrix that connected wetland types with key ecological functions—such as water retention, nutrient transformation, carbon sequestration, sediment stabilization, and habitat provision. Wetland polygons were ranked by functional performance and potential restoration value using terrain models, hydric soils Soil Survey Geographic Database (SSURGO), and historical imagery.

The result was a functional wetland geodatabase integrating data from National Agriculture Imagery Program (NAIP) imagery (2009), US Geological Survey (USGS) hydrography, and Google Earth time series, supporting:

- Wetland restoration and mitigation planning
- Rapid assessment calibration
- Environmental permitting and Clean Water Act reporting.

This approach created a replicable, AI-ready geospatial framework that demonstrated how remote sensing and multi-scale analysis can enhance data-driven wetland management. For developing regions like Vietnam, the Canadian River model highlights the importance of multi-tiered assessments, interoperable geospatial databases, integrated hydrological–ecological planning and community participation. These lessons directly inform the AI-based Integrated Wetland Connectivity Framework proposed for Song Dam Lake, Da Nang, ensuring scientific consistency and global applicability.

London Wetland Centre: Barn Elms, UK

The London Wetland Centre (LWC) at Barn Elms, southwest London (LWC 2025) represents one of the most successful examples of large-scale urban wetland restoration and management in Europe. Developed by the Wildfowl and Wetlands Trust (WWT) in partnership with the Environment Agency (EA) and the London Borough Council, the project transformed a 130-hectare disused reservoir complex into a functioning artificial wetland reserve with rich ecological, educational and cultural values.

At the core of this model lies the principle of ‘Design with Nature’, integrating ecological processes, landscape planning and hydrological functions. The site design mimics the natural floodplain dynamics of the River Thames through a mosaic of interconnected habitats, including:

- Deep lagoons: Maintaining water depth and flow suitable for diving bird species such as kingfishers and pochards.
- Shallow marshes and mudflats: Providing foraging grounds for waders and migratory birds.
- Reedbeds and wet meadows: Enhancing nutrient retention, carbon sequestration, and water purification.
- Ecological corridors and natural regeneration zones: Linking the site with the wider Thames green network to support wildlife connectivity.

Environmental management at LWC is supported by smart monitoring technologies, including IoT-based water sensors, GIS mapping, and Unmanned aerial vehicle (UAV) imagery, which continuously track water levels, quality, vegetation dynamics and avian activities. All data are stored within the Wetland Information Management System (WIMS), providing a basis for scientific research, public education and adaptive management.

Over more than two decades, the London Wetland Centre has demonstrated measurable ecological and social outcomes:

- Restoration of over 30 wetland habitat types, supporting more than 180 bird species, including rare species such as bittern (Botaurus stellaris) and common snipe (*Gallinago gallinago*).
- Improved local water quality, with nitrogen and phosphorus concentrations reduced by over 60% compared to pre-restoration conditions.
- Development into Europe's leading urban wetland education and ecotourism hub, attracting over 200,000 visitors annually, including 20% school and university participants engaged in 'Wetland Explorers' and 'Citizen Science' programs.

The LWC illustrates how urbanization and wetland conservation can coexist through ecosystem-based planning, inclusive governance and science–policy–community integration. Its holistic framework aligns ecological restoration with socio-cultural values and long-term sustainability.

For developing countries like Vietnam, the London Wetland Centre offers valuable insights for creating artificial or hybrid wetland reserves that combine education, ecotourism and research functions. These lessons provide strategic guidance for the proposed AI-assisted Urban Wetland Conservation and Education Model at Song Dam Lake (Da Nang)—integrating AI, IoT and remote sensing to support real-time monitoring, habitat restoration and community-based environmental education.

Comparative Insights and Lessons for Vietnam

The Canadian River Watershed (USA) and London Wetland Centre (UK) represent two complementary wetland management models: one emphasizes landscape-scale functional assessment, while the other focuses on urban ecological restoration and education. Together, they offer valuable guidance for developing AI-assisted and community-based wetland frameworks in Vietnam.

First, the Canadian River model demonstrates the value of multi-tiered, data-driven assessment, integrating GIS, HGM indicators, and functional evaluation to quantify ecosystem services. In contrast, the London model translates ecological processes into landscape design, applying ecosystem-based planning to restore wetlands within urban settings. For Vietnam, combining these two approaches can ensure that wetland restoration is both scientifically grounded and ecologically functional.

Second, both projects highlight the role of technology and interoperability. The Canadian River project established a geodatabase integrating remote sensing and hydrological data, while the London Wetland Centre uses IoT sensors, UAVs, and GIS-based management systems. Applying these technologies to Song Dam Lake will enable real-time environmental monitoring, predictive modeling, and adaptive management.

Third, governance and community engagement are essential. The Canadian River model relies on interagency cooperation, while the London model promotes public–private–community partnerships. Similarly, Vietnam should foster collaboration among local authorities, universities, NGOs and citizen scientists, combining conservation with environmental education and eco-cultural tourism.

Integrating these lessons, the AI-based Integrated Wetland Connectivity Framework proposed for Song Dam Lake will merge the following: multi-tiered ecological assessment, smart monitoring and data sharing, and community participation and education. This integrated approach will create a digital and ecological twin of the Song Dam wetland, linking science, technology and culture to enhance biodiversity conservation, climate-responsiveness, local resilience and sustainable eco-tourism.

References

Canter LW (1977) Environmental impact assessment. McGraw Hill, New York, p 175, 241

Climate Action News (2025) Personal service email

France RL (2016) From Barn Elms reservoirs to the London Wetland Centre. In: Environmental restoration and design for recreation and ecotourism. CRC Press, pp 125–212

Graeme L, others (2010) Connectivity conservation management: a global guide. Earthscan

Guterres A (2022) Message on world meteorological day

IDB (2018) Disaster and climate change: risk assessment methodology for IDB projects, p 13

Kim KG (2017) Development of an integrated smart water grid model as a portfolio of climate-smart cities. J Smart Cities 3(1):25–32

London Wetland Center (2025) website

Lin P, Lu Q, Li D, Chen Y, Zou Z, Jiang S (2019) Artificial intelligence classification of wetland vegetation morphology based on deep convolutional neural networks. Nat Resour Model 33(1). https://doi.org/10.1111/nrm.12248

Nundloll V, Porter B, Blair GS, Emmett B, Cosby J, Jones DL, Chadwick D, Winterbourn JB, Beattie P, Dean G, Shaw R, Shelley W, Brown M, Ullah I (2019) The design and deployment of an end-to-end IoT infrastructure for the natural environment. Future Internet 11(6), 129. https://doi.org/10.3390/fi11060129

Saint Mary's University of Minnesota, Geospatial Services (2013) A landscape-level approach to wetland functional assessment: Wetland mapping & functional assessment, Canadian River Watershed, New Mexico [Technical report]. Saint Mary's University of Minnesota

Schwab K (2016) The fourth industrial revolution. Penguin Random House

Sin HS (2025) Generative AI trends and AI agent, paper presented at google startup campus, Seoul, Korea, p 15

Tiner RW (2003) Geographically isolated wetlands of the United States. Wetlands 23(3):494–516. https://doi.org/10.1672/02775212(2003)023[0494:GIWOTU]2.0.CO;2

Chapter 6
A Practical Illustration of AI in Prediction and Optimization Processes: Delivery of End-to-End Solutions

Kwi-Gon Kim, Ho-Gil Park, and Bong-Koo Lim

Abstract As climate action turns towards implementation of ambitious action with AI, cities and regions are ready to show solutions. This chapter demonstrates the steps associated with prediction and optimization of climate solutions in cities including data collection, algorithm development and solution suggestions. These steps can be taken either in real situations or in the virtual world (digital twin), and aim to provide more value and possibilities for cities.

Keywords Urban sensing · Virtual space · Algorithm · IoT sensors · Data storage · AI hardware · AI chip · Perplexity · AI Infra · GPU · AI data center · Data platform · Spheres · REfik Anadol · Nubison AIOT · LCA application

6.1 Implementation Process

The city climate planning process and its components have led to the development of new terminology. There are many different versions of the process that use AI with an urban project. The author views the process culminating in climate solutions as consisting of six AI technical components: (1) urban sensing, (2) data transformation, (3) data storage and cloud-computing platforms for data management, (4) data platform, (5) data analytics and (6) optimizing models in a practical manner (Fig. 6.1).

K.-G. Kim (✉)
Seoul National University, Seoul, South Korea
e-mail: kwigon@snu.ac.kr

H.-G. Park
ArkLABS, Seoul, South Korea
e-mail: wonja720@daum.net

B.-K. Lim
Asia Environment Institute, Seoul, South Korea
e-mail: ecotourism@naver.com

K.-G. Kim (ed.), *AI Transformation of Cities for Climate Solutions*,
https://doi.org/10.1007/978-3-032-21940-4_6

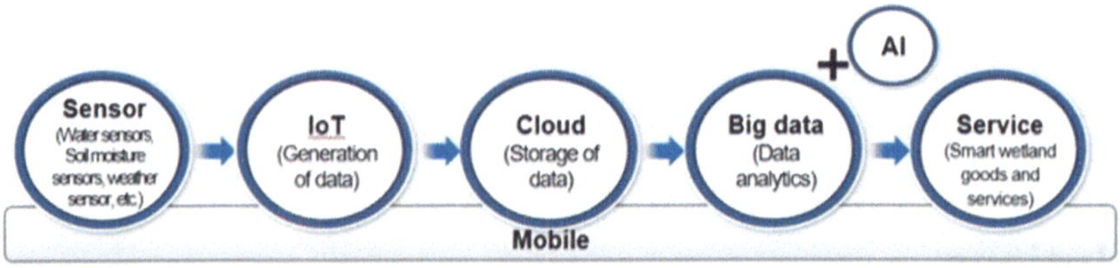

Fig. 6.1 Processes that use AI with an urban project

Fig. 6.2 Six activities of the AI process (Snowflake World Tour 2025)

Box 6.1

Standard Paradigm Shift (Schwab 2016):

We are at the beginning of a revolution that is fundamentally changing the way we live, work, and relate to one another. In its scale, scope and complexity, what I consider to be the fourth industrial revolution is unlike anything humankind has experienced before.

Source: Schwab, K., *The Fourth Industrial Revolution*, Penguin Random House, UK, 2016, p. 1

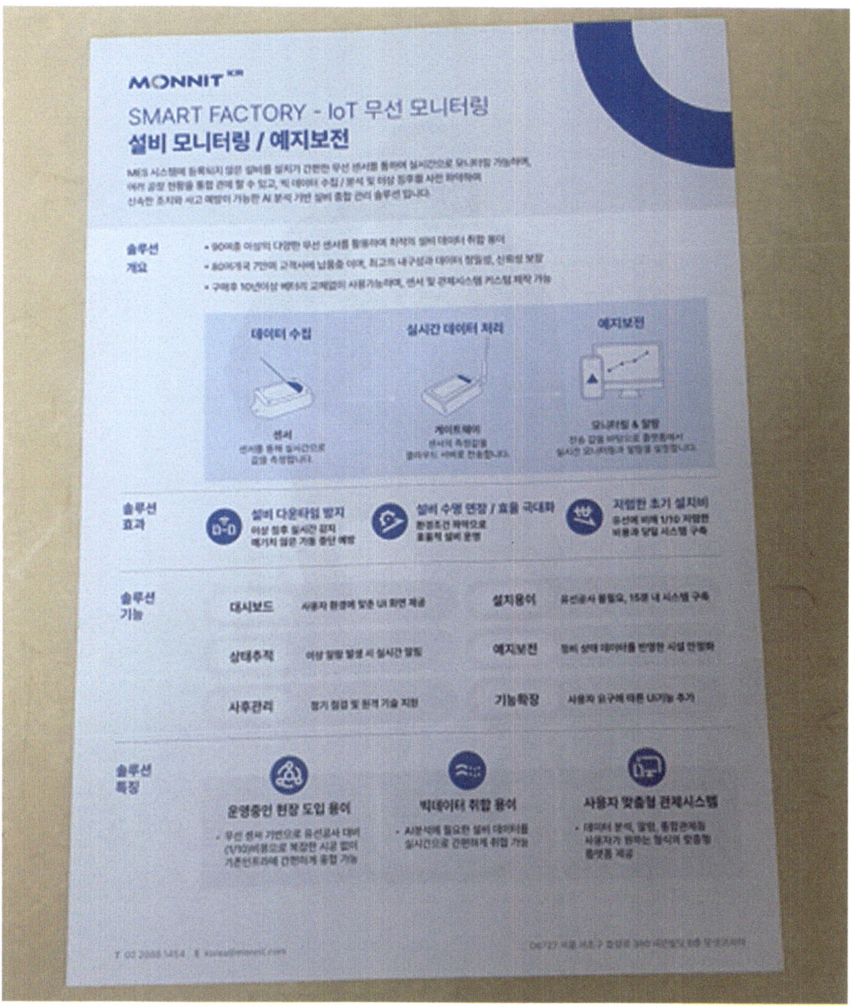

Fig. 6.3 Data collection, Internet of Things (IoT) transmission of measured data to cloud server, and platform for real-time monitoring and alarm set-up. *Source* MONNIT Korea, 2025

The diagram in Fig. 6.1 extends the urban planning and design approach to include use of digital innovation for data collection, data transformation, data storage, data reduction, data mining, data analysis, AI modeling (statistical and visual) and customer service.

The Snowflake World Tour (2025) has identified six activities of the AI process. These steps include: (1) AI data platform, (2) data engineering, (3) analytics, (4) AI intelligence, (5) application and collaboration, and (6) next steps: Learn, Build, Grow.

Figure 6.2 shows these six activities in order.

Figure 6.3 shows another version of the AI process for the smart factory. This includes data collection, IoT transmission of measured data to a cloud server, and a platform for real-time monitoring and alarm set-up.

Fig. 6.4 Collection of data and storage using a sensor and server for groundwater

Fig. 6.5 Different types of air quality sensors. *Source* SMARTLINE Co., Ltd., 2025

Fig. 6.6 Building crack gauge sensor. *Source* arkLABS Co., Ltd., 2025

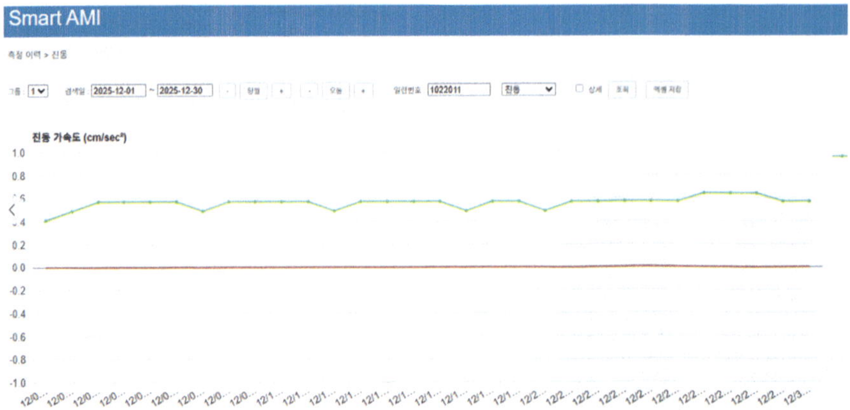

Fig. 6.7 Vibration acceleration

There are several basics or prerequisites that must be fulfilled for appropriate use of AI. These basics are as follows:

- Types of sensors
- Data transformation
- Virtual space for data storage
- Big data platform
- AI platform
- Knowledge of algorithms

6.2 Urban Sensing: Sensor Networks

The evolution of urban AI science and computing technology has generated new functions that enrich the extension of urban sensing. This section presents representative applications of sensors in a series of pictorial formats.

Pictorial illustrations of different types of sensors for data collection are shown in Figs. 6.4, 6.5 and 6.6.

Real-time data monitoring and analysis of crack measurement are shown in Figs. 6.7, 6.8, 6.9, 6.10, and 6.11:

6.3 Data Transformation

See Figs. 6.15 and 6.16.

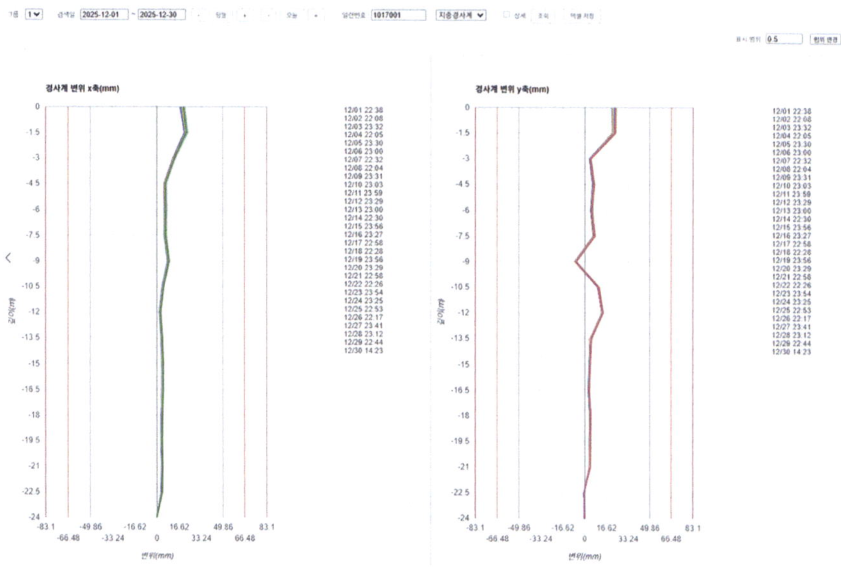

Fig. 6.8 Inclinometer

Fig. 6.9 Crack

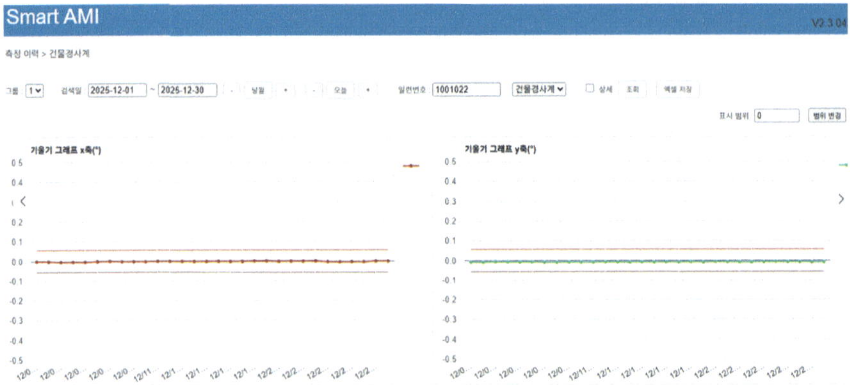

Fig. 6.10 Inclination

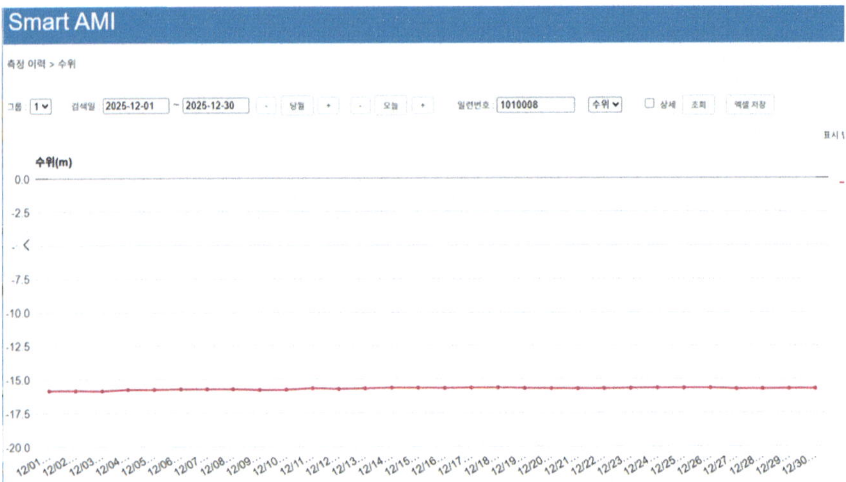

Fig. 6.11 Water gauge

6.4 Data Storage and Cloud-Computing Platforms for Data Management

AI Infra includes storage for the integration of data from diverse sources with cloud AI service such as Google Cloud AI service and computational infrastructure. AI hardware involves the data center (computing capacity), AI chip (GPU, TPU, etc.) and memory, and AI software includes ChatGPT, Google Gemini and Perplexity. Data centers require the greatest application of technology (Figs. 6.17, 6.18, 6.19, 6.20, 6.21, 6.22, 6.23).

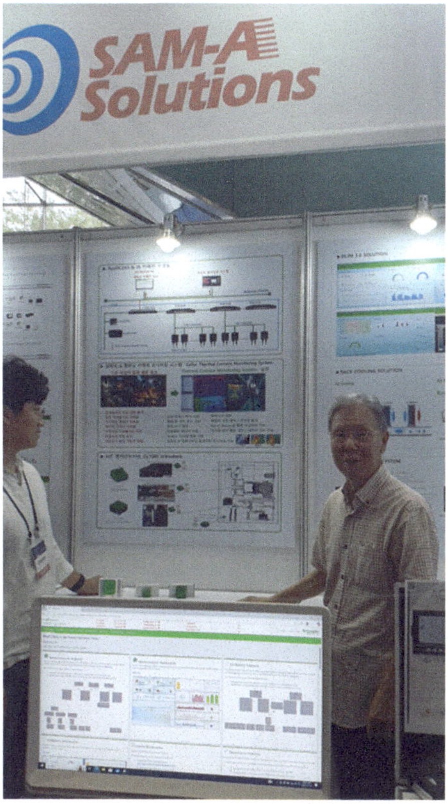

Fig. 6.12 IoT sensors (green colour) sending data to server without wire. *Source* SAM-A Solutions, 2025

Fig. 6.13 Different uses of sensors (maritime gas emission monitoring solutions, ambient air quality monitoring solutions, stack emission monitoring solutions, etc.). *Source* SNARTLINE, 2025

Fig. 6.14 Methane emission monitoring solutions. *Source* SMARTLINE, 2025

6.5 Data Platform

With the rise of ML and AI, as well as Geographic Information Systems (GIS), the applications of big data analytics in the context of AI city planning have diversified,covering a wide range of applications starting with climate change, environmental monitoring, public safety and adjusting power distribution based on consumption patterns, and enhancing urban living by providing personalized and efficient solutions to everyday life style with personalized AI technology.

A data platform is a comprehensive end-to-end solution for all the data. The platform includes data, storage, computing environment, data analytics and applications (Figs. 6.24, 6.25).

Fig. 6.15 IoT design review gate process. *Source* ASUS, 2025

6.6 Data Analytics: AI Algorithms Including Regression for Prediction

6.6.1 Data Mining and Other Technical Terms

Figure 6.26 illustrates the connection between data mining and other technical terms including machine learning, AI, data management (databases), big data and cloud computing to mine knowledge from data (Zheng 2018).

Fig. 6.16 AI-based U-NIS PRO: Integrated IoT networking systems of buildings (apartments, houses, etc.) connected with AI agents. UbiQuoss (2025)

6.6.2 *Incorporating AI into the Urban Planning and Development Process: AI Models and Algorithms*

- Deep learning workflow
- Deep learning accelerator platform
- Spatiotemporal data transformation for deep learning
- AI models platform (Figs. 6.27, 6.28, 6.29, 6.30).

6.7 Optimizing Models as a Planning Tool

Using AI models with training data helps to identify the best policy for maximizing sustainability objectives.

Fig. 6.17 Amazon Web Services (AWS) supercomputer (or AI server)

Fig. 6.18 Cutting edge AI Infra for liquid cooling systems, AI accelator, AI memory, AI DC SecureEdge, AI cloud manager, etc.

6.7.1 Learning Data Setup

One of the best approaches to create AI models and algorithms is to have knowledge of the types of AI associated with given types of projects or studies. For example, in ecological prediction and optimizing modeling, it is necessary to develop a learning dataset from the entire process of traditional ecosystem evaluation shown on the left-hand side of Fig. 6.31.

Fig. 6.19 AI PMDC: AI data center with a cooling system. *Note* The blue building shows an AI data center with 256 GPUs (32 servers multiplied by 8 GPUs). The remaining buildings are for electricity supply

In addition, it is required to consider principles of realizing AI agents' potential for achieving the goals of modeling including fairness, inclusivity, transparency and accountability derived from knowledge of humans, nature and algorithm study. Specific principles and guidelines may be needed for guiding the deployment of AI agents to facilitate responsible advancements towards the Sustainable Development Goals.

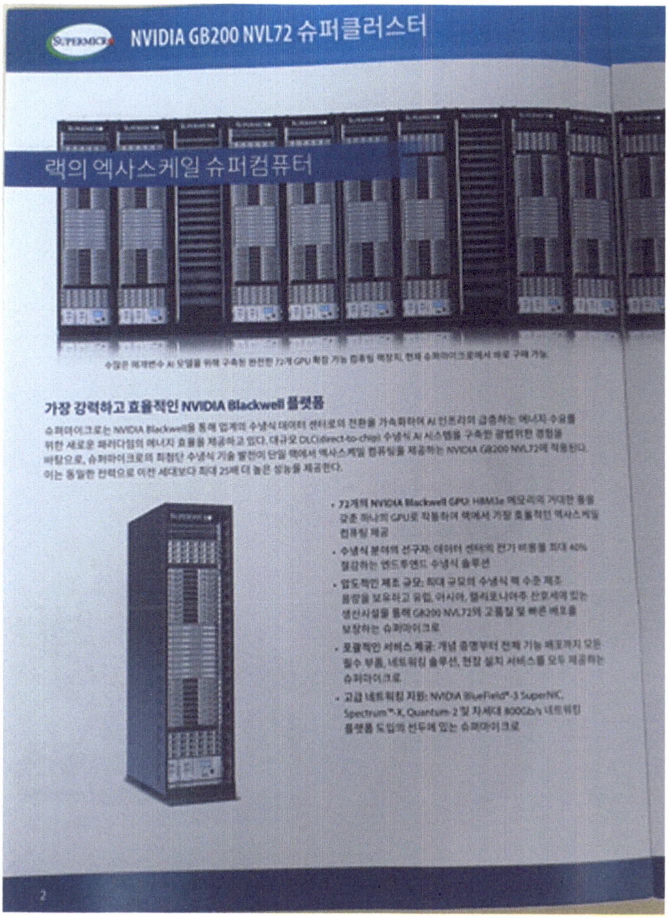

Fig. 6.20 NDIVIA Supercomputer: Blackwell platform. *Source* SUPERMICR, 2025

6.7.2 *Understanding Algorithms*

AI algorithms promote innovation and drive the development of intelligent city science and technology towards sustainable development.

AI algorithms are sets of instructions that enable machines to perform tasks that typically require human intelligence. These algorithms analyze data, identify patterns and make decisions based on learned information. They are the core components that drive AI systems, allowing them to perform functions like recognizing bird species.

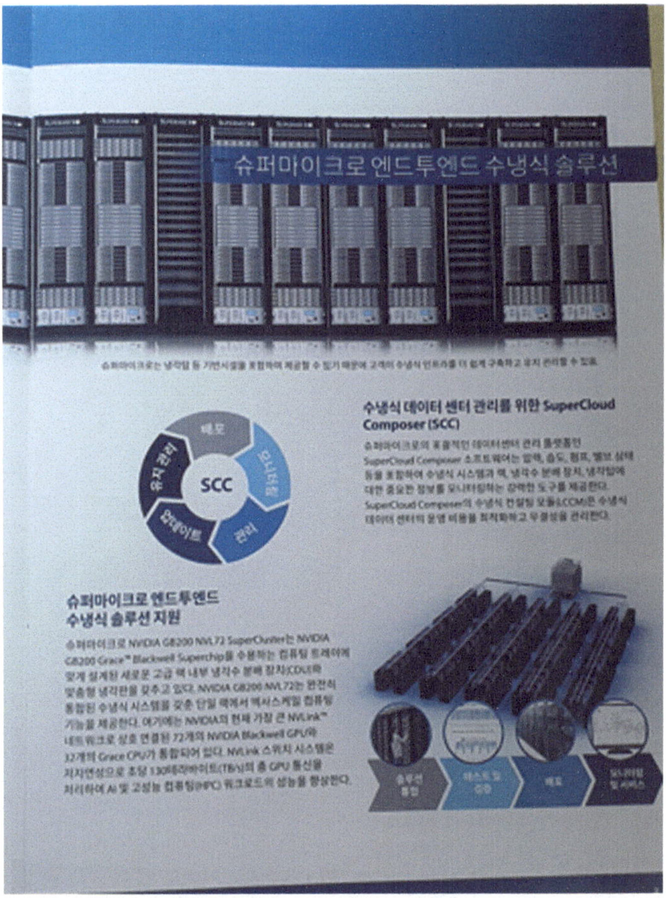

Fig. 6.21 AI data center cooling solution: SuperCloud computer (SCC). *Source* SUPERMICR, 2025

Here's a more detailed overview of the main functions of algorithms:

- Process information: AI algorithms act as the 'brains' of AI systems, processing input data to understand its meaning and identify relevant information.
- Learn from data: Algorithms learn from examples, often labeled data, to recognize patterns and relationships.
- Make predictions: Based on learned patterns, AI algorithms can predict future outcomes or classify new data points.
- Automate tasks: They enable machines to perform tasks autonomously, without explicit human direction.

Fig. 6.22 2025 AI data center Summit Korea, COEX, Seoul, Korea: A to Z for Sustainable Data Center Industry

Types of AI algorithms and examples include:

- Supervised learning: Algorithms learn from labeled data, where both input and desired output are provided. Examples include linear regression, decision trees and support vector machines.
- Unsupervised learning: Algorithms learn from unlabeled data, discovering patterns and structures without explicit guidance. Examples include clustering algorithms (like K-means) and dimensionality reduction techniques.
- Reinforcement learning: Algorithms learn through trial and error, receiving rewards or penalties for their actions. Examples include Q-learning and deep Q-networks.

Fig. 6.23 Database performance management solution: Sherpa AI associated with ORACLE and MongoDB. *Source* Sherpasoft, 2025

In essence, AI algorithms are the fundamental building blocks of artificial intelligence, enabling machines to learn, adapt and perform complex tasks that were once exclusive to humans (ChatGPT 2025).

6.7.3 Application Areas in Cities: AI for Advancing Urban Systems

Amplified migration patterns, built-up land expansion, infrastructural changes and urban sprawl are driving and responding to urban climate changes. The unprecedented scale and rate of these changes and their climate impacts call for transformative approaches to understand urban systems and highlight solutions addressing multidimensional tradeoffs and synergies. State-of-the-art artificial intelligence (AI) approaches highlight tremendous potential to transform urban systems science through advancements in novel measurement, modeling and simulation, and decision support systems for urban climate solutions.

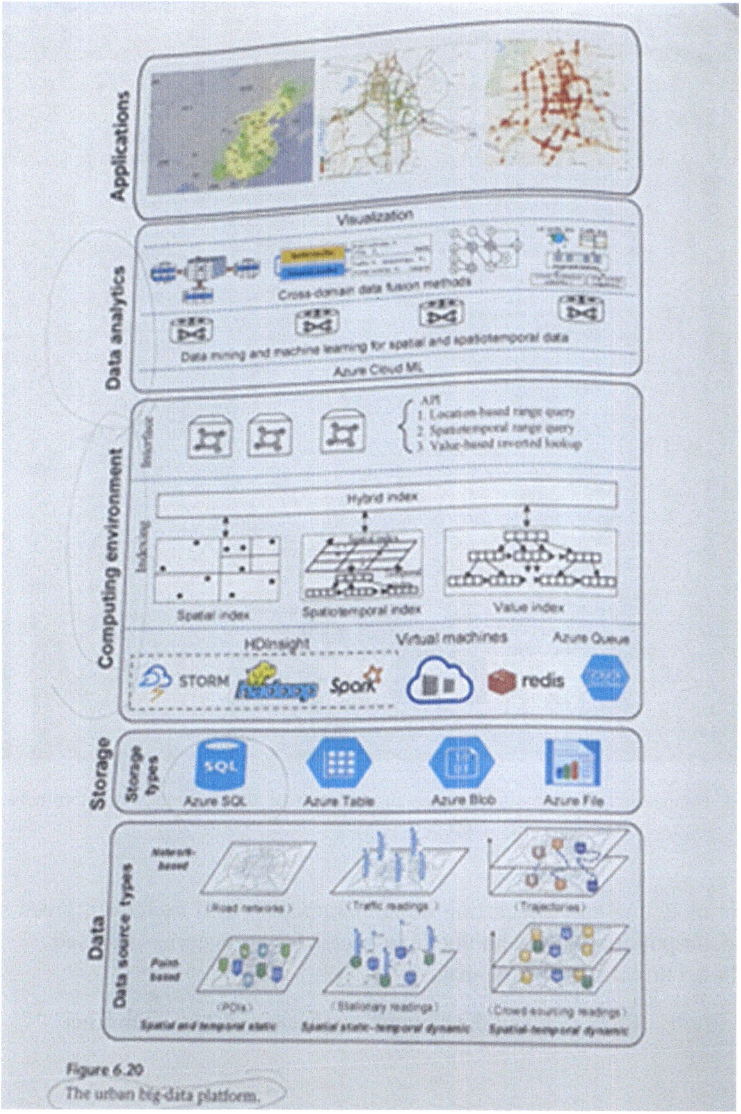

Fig. 6.24 The urban big data platform. *Source* Yu Zheng, Urban Computing, 2018, p. 302

Algorithm-related topics include:

- Multimodal foundational models for downscaled urban climate
- AI algorithms for urban climate modeling making long-term climate prediction
- GHG accounting
- AI for urban sensing and measurements
- AI-based urban climate emulators

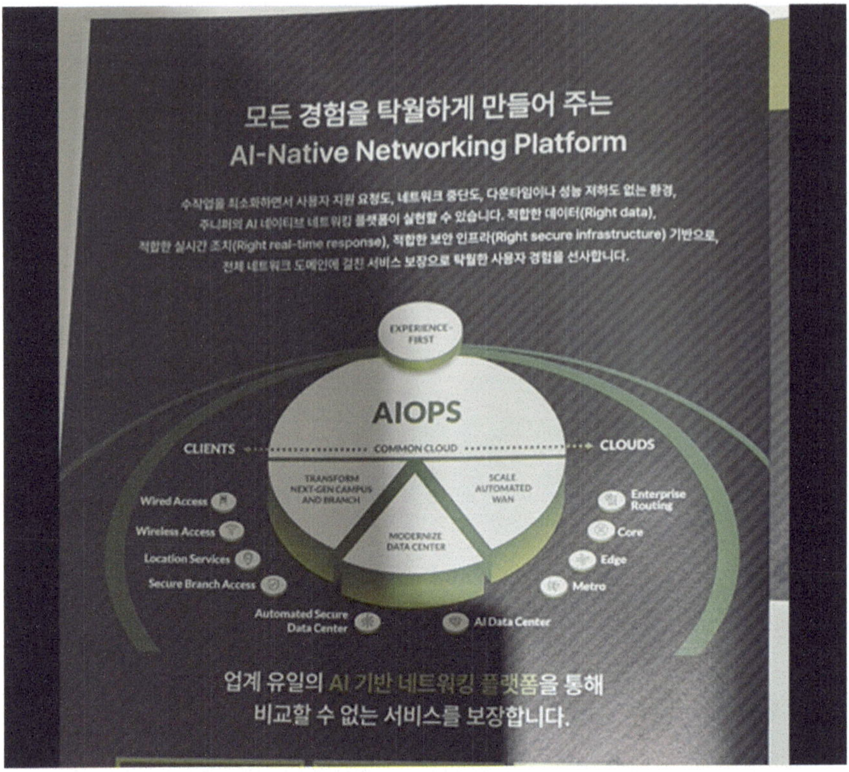

Fig. 6.25 Connection of clients and clouds in modernizing data centers: AI-native networking platform. *Source* Juniper networks, 2025

- AI in modeling and forecasting urban parameters: land use/cover, green space, wind, air pollution, heat and flooding; energy/water/materials demand
- AI-based urban complexity analysis.

The details of research and development topics are discussed in Chap. 17.

6.7.4 *Case Study Examples of Prediction and Optimization Algorithms*

Amazon's Headquarters: The Spheres (Korean Air 2025a)

This is a place where humans and algorithms share the work of managing perfectly controlled environment, and where nature is aided by cutting-edge technology.

The Spheres is an ecosystem designed by algorithms to provide:

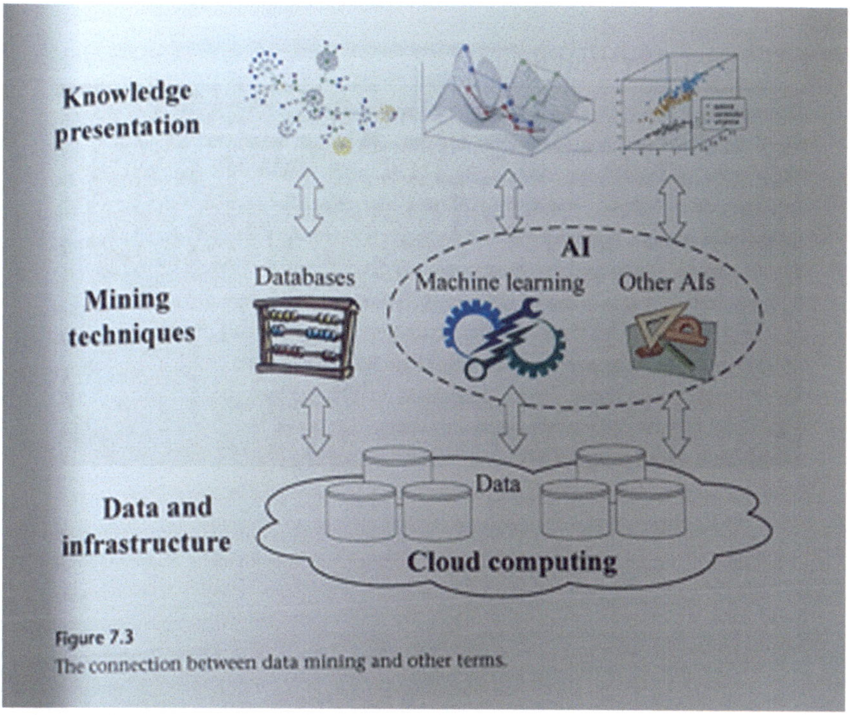

Fig. 6.26 The connection between data mining and other terms. This system includes data infrastructure (data and cloud computing), mining techniques (databases, ML and other AIs), and knowledge presentation. *Source* Yu Zheng, Urban Computing, 2018, p. 313

- Digital generation
- ‘Work-life harmony’ which holds that working in a natural setting makes people more creative and resilient
- A laboratory that tears down the boundaries between work and life
- A restorative ecosystem that helps employees activate their senses and regain their rhythm.
- Cutting-edge technologies to aid nature
- Plants that thrive in a perfectly controlled environment where a consistent cycle of day and night is maintained
- A high-tech ventilation system which ensures that fresh air continuously circulates through spaces
- Management of the environment shared by humans and algorithms
- Campuses designed to resemble forests to explore ways to integrate nature into technology.

(Source: Korean Air Inflight Magazine, July/August/2025, p. 22).

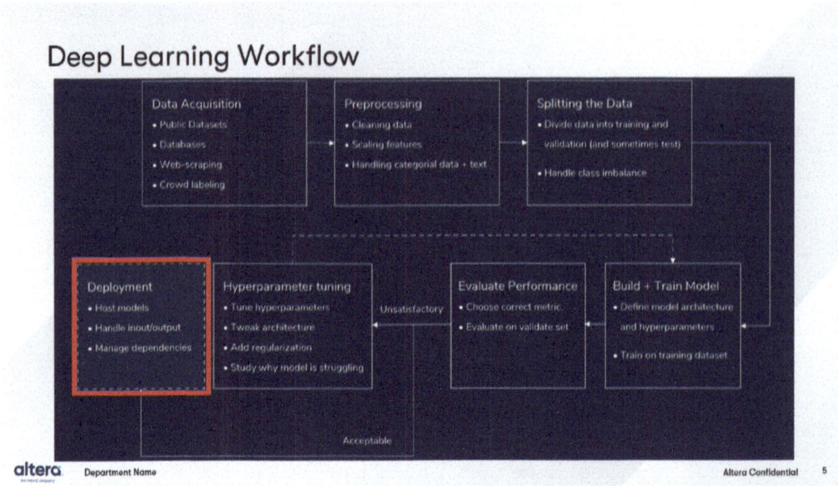

Fig. 6.27 Deep learning workflow. This workflow involves data acquisition, processing, splitting the data, building + training model, evaluating performance and deployment. *Source* UNIQUEST, 2025

Fig. 6.28 NDVIA-based deep learning acceleration platform (DLAP): Expanding AI frontiers in manufacturing, retail and beyond. *Source* ADLINK, 2025

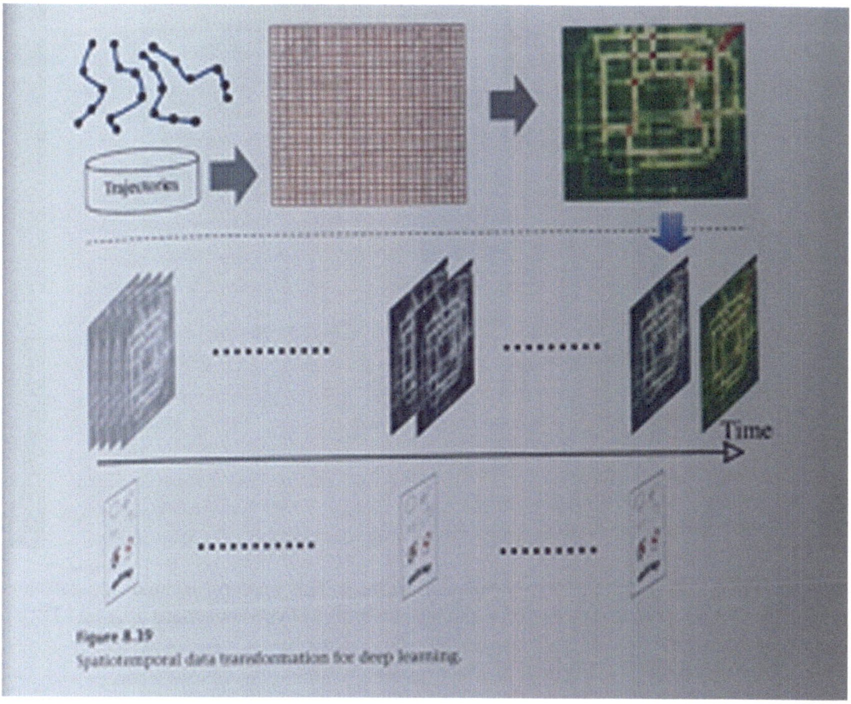

Fig. 6.29 Spatiotemporal data transformation for deep learning. *Source* Yu Zheng, Urban Computing, 2018, p. 469

The Poet of Algorithms: Refik Anadol (2025b)

Refik Anadol is a world-renowned artist and a vanguard in the aesthetics of data and machine intelligence. The major features of his AI work are as follows:

- Dissolving the very boundaries between technology and art
- Actualizing human-centered values through the compelling medium of AI
- Redefining the frontiers of AI and art: If machine can imagine and dream, to what horizons can art then expand?
- Reflecting real-time climate data onto Gaudi's iconic building in Barcelona
- Reconstructing and visualizing emotions, memories, ecologies and civilizations through data and algorithms
- Creating new forms of beauty where humanity, nature, machines and memory intersect through AI that is trained on data imbued with cultural, ecological and emotional memory.

(Source: Korean Air Inflight Magazine, JULY/August/2025, pp. 98–100).

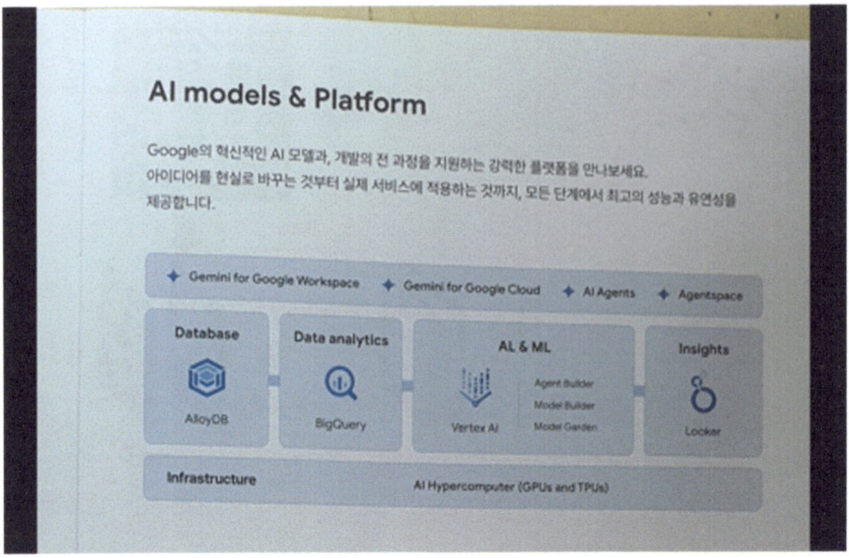

Fig. 6.30 The whole process of AI models and platform. This platform includes the database, data analytics (Bigquery), Vertex AI (ML), and insights with AI hypercomputer(GPUs and TPUs). *Source* Google, 2025

Implications for the Song Dam Lake Case Study

- Expanding the role AI plays in cities
- Creating new types of eco-cultural tourism where water, ecology, climate and community livelihood are connected and sustainably optimized through AI that is trained on data imbued with water, habitat suitability, carbon sequestration and community participation
- Visualizing images of project activities through data and algorithms
- Redefining the frontiers of AI and planning
- A living laboratory that tears down the boundaries between human and machine intelligence
- Data collection that is aided by cutting-edge technologies including AI
- Humans and AI algorithms sharing the work of managing the climate and environment
- Use of sustainability indicators for AI algorithms.

We need to connect cities and their community people, visitors and other living things through data-driven AI algorithms. We have to work with datasets like wind, green and water infrastructure, wildlife habitat suitability, climate patterns and carbon cycle to train multi-AI agent models for new forms of urbanism.

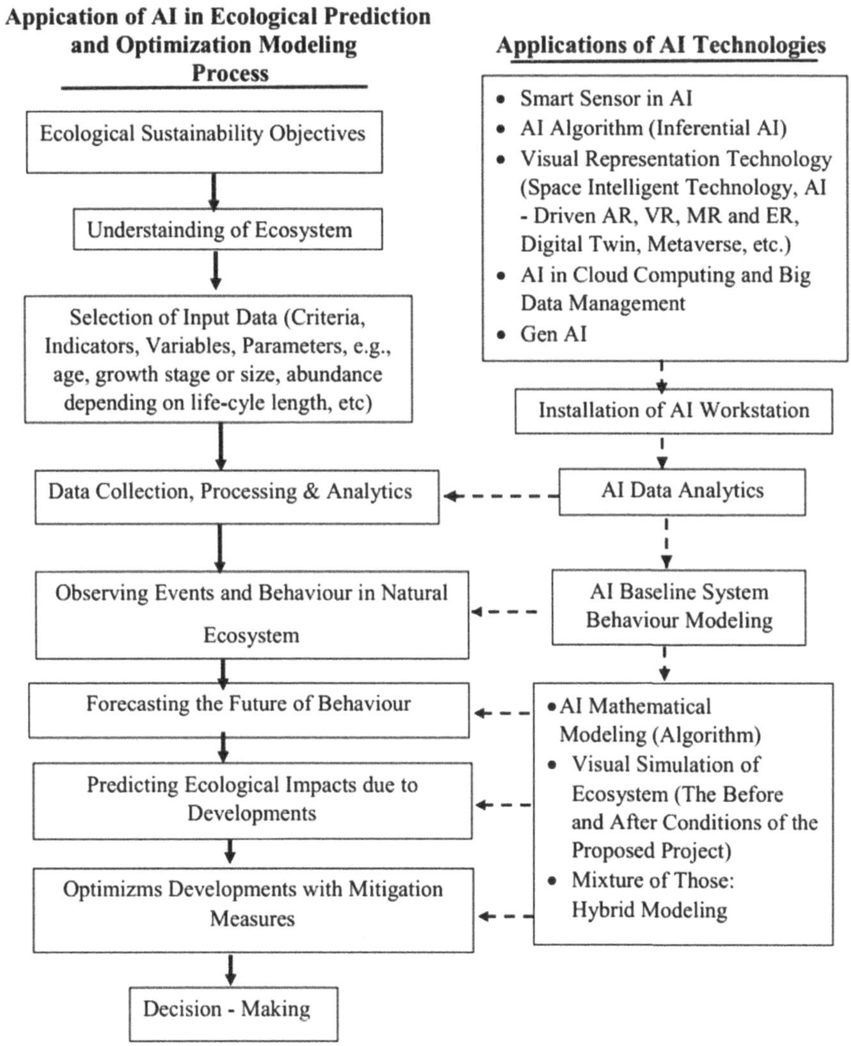

Fig. 6.31 Application of AI in ecological prediction and optimization modeling

Power Demand Forecast Based on Power Usage-Related Data Analytics, NUBISON AIOT

This algorithm uses regression called "LPCP (life-style based power consumption prediction)" model with 0.91 of algorithm accuracy (Fig. 6.32).

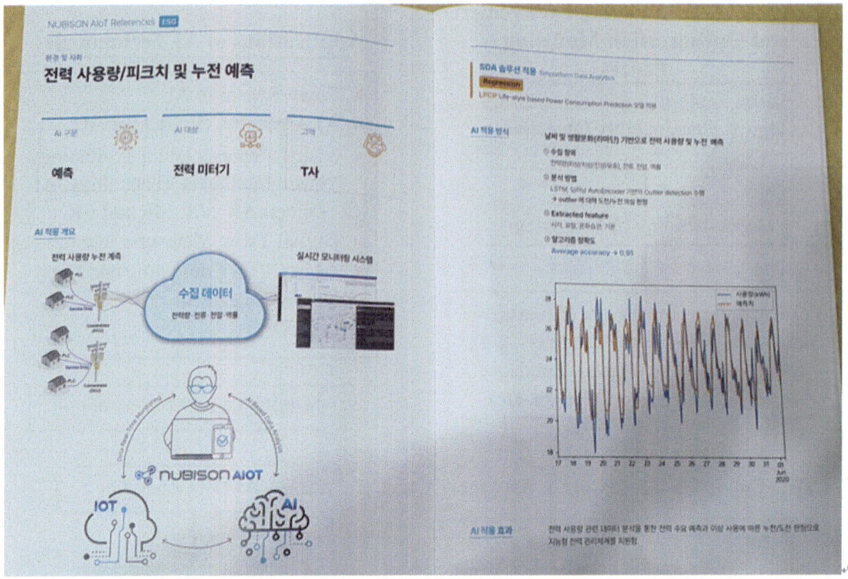

Fig. 6.32 An illustration of the electricity demand prediction algorithm to support intelligent electricity management system. *Source* NUBISON AIOT, 2025

LCA Application

AI creative disruption technologies over the life cycle of project activities and potential usage of AI technologies are crucial in eco-cultural sustainable tourism development in the Song Dam Lake and its surrounding area, Da Nang City, Vietnam. The author developed potential AI technologies to be used for the life cycle stages of all project activities, from the planning and design to the capacity building and education stages. Table 6.1 provides a short list for illustration purposes. It could also include climate impact assessment and sustainability cost-benefit analysis in the planning process.

Table 6.1 List of AI creative disruption technologies for life cycle stages of project activities with emerging startups

Life cycle stage of project activities	Application areas	AI technologies (AI framework)	Best practices (Startups)
1. Planning and design stage	**1.1. Forecasting models for developing future planning indicators** • Eco-cultural tourism demand modeling • Eco-system change modeling • Future climate change modeling: Carbon footprint modeling • Disaster early warning modeling	• AI forecasting algorithms (e.g., statistical and regression analysis) • AI geospatial portfolio planning technology • Demand estimation with AI and GIS application software • Optimization for decision making	**1. Leisure demand forecasting** [Component: AI demand forecasting, smart city] 0. Startup: HeBAN, Korea
	1.2. Virtual presentation of eco-cultural tourism site images • 3D printing /3D imaging before and after development	• Visual presentation of future images – Drone image embedded with hyper-spectral sensor – VR, AR, MR, ER – Digital twin – Metaverse • Geographic information system and GPS for digital ecological, environmental and hydrological mapping – Design software: AutoCAD, MAX, Sketchup, Lumion, etc	**2. AI Agent** [Component: Digital transformation solution, data visualization and analytics, future -oriented solutions] 0. Startup: BIZENTRO, Korea
	1.3. Site data collection and analytics • Big data storage and analytics • Data management	• Survey equipment (GPS survey, optical survey (GEO) theodolite, total station, etc.) • AI survey on needs, challenges and solutions for eco-cultural tourism • Big data platform • AI platform • Computer-assisted software (e.g., SPSS, ArcGIS, Mapinfo) • AI data and machine learning	**3. Platforms for the AI era** [Component: AI data smart platform connection and integration] • Startup: METABUILD, Korea
	1.4. Suitability, risk and carrying capacity analyses Layer solutions GIS system connected with drones	• GIS and remote sensing tools • ArcGIS package • Digital shadow models showing layers of the physical world and digital shadow (Source: BMI-Lab)	**4. Satellite metadata innovation** [Component: AI-powered image analysis solutions, SAR/EO-based satellite images, visualization of the spatial and location information] • Startup: Stellar vision, Korea

(continued)

Table 6.1 (continued)

Life cycle stage of project activities	Application areas	AI technologies (AI framework)	Best practices (Startups)
	1.5. Master plan preparation 1.5.1. Land-use zoning • GIS mapping-based zoning • Land-use regulations at the national and provincial levels • Growth management at the city level	• AI technology for connecting with the community and city • AI technology for connecting across the tourism site	**5. Land use: integration and connection** Powered by Aisop [Component: Physical - digital integration and connection of land use through a miriad of AI sensors working 24 h/day] • Startup: Rex
	1.5.2. Activity space programming • Area size by land use • Spatial magnitude by tourism activity	• Spatial intelligent technology	**6. Air quality and water quality-based space analysis solution** [Component: Harmful gas sensing and notification system] • Enterprise and startup: HYUNDAI Engineering and Construction, Piquent, Korea
	1.5.3. Facily programming • Water-based recreation facilities • Tourism furniture and amenities • Media art center: Art and nature in harmony	• Chat GPT nature interpretation devices • Smart kiosk information zone • AI furniture and lighting • Wireless CCTV camera watch • AI computing center • AI bike count and rack	**7. Generative AI. made fast, efficient and affordable hardware** [Component: Hyper accelerated hardware solutions for emerging AI applications which provide innovative and practical AI hardware] • Startup: HYPER ACCEL, Korea
	1.6. Sectoral design 1.6.1. Landscape and ecosystem • Wetland restoration / rehabilitation • Blue-green area network • Enhancing the connectivity of urban wetlands • Hardscape • Picnic and camping sites Community garden	• Integrated observations and modeling of GHG budgets at the ecosystem levels • AI smart farming technology • AI food lab • Certification systems for sustainable practices of green roofs and green walls • Near-infrared photography to determine health of plants in terms photosynthesis: Near-infrared cameras, digital camera and post-processing software	**8. Vegetation showing up as red in the near—infrared photograph of Fairbanks, Alaska, USA** • Institution: Sandia National Laboratories, USA **9. Nature-based solutions (NBS)** through a robust and scalable measurement

(continued)

Table 6.1 (continued)

Life cycle stage of project activities	Application areas	AI technologies (AI framework)	Best practices (Startups)
			Reporting, and verification (MRV) system [Component: AI, satellite data, and smart sensor to accurately measure and monitor carbon sequestration and emissions] • Company: Transitry, Singapore
	1.6.2. Building • Operation and maintenance center • Restaurants • Observation tower • Central information center	• Smart IoT building • Digital botanic garden • Digital locker • Green building technology • Building integrated photovoltaics • Digital meters for gas, water and electricity • Smart LED signage for building	**10. Eco-friendly smart building living water purification plant** [Component: Data driven building water quality management using data on usage, consumption habits, carbon emissions, and remote, integrated monitoring system (TMS)] • **Startup: GEOGRID, Korea** **11. Wavy room for buildings** [Component: Building design, RandD, and production of modular homes, powered by a proprietary automation solution that combines AI with advanced building software, BIM] • Company: Space wavy, Korea
	1.6.3. Circulation • Integrated smart circulation System • Buggy services	• AI parking lot and management • AI vehicle information terminal system	**12. AI-based smart parking control technology** [Component automatic: AI CCTV, parking time, toll collection] • Startup: LIGHTVISION, Korea

(continued)

Table 6.1 (continued)

Life cycle stage of project activities	Application areas	AI technologies (AI framework)	Best practices (Startups)
	1.6.4. Energy • Enhancing the site's energy efficiency • Smart energy micro -grids	• ESS • Smart solar collector field • Solar on–off micro-grids – Solar site lighting systems – Solar streetlights – Solar water pump – Solar water purifier – Solar cooking systems – Solar heating and water heating -Solar drier – Solar water heater – Solar sculture – Solar playground – Solar deck (boardwalk) and bench – Solar community micro-grids systems	**13. AI-based energy optimum management solution** [Component: Energy consumption analysis algorithm, energy management platform] • Company and startup: HYUNDAI Engineering and Construction, Green Data, Korea
	1.6.5. Water • Reducing the site's waste production • Water recycling • Integrated water resource management • Smart water grids	• Tele monitoring system (TMS) • Rainwater harvesting technology • Rainwater supply meters • Smart water controller • Water recycling technology	**14. AI water management (water round)** [Component: Watershed water management urban flooding, decentralized water, treatment, intelligent water quality sensor] • Corporation: Kwater, Korea **15. AI water surface quality management** (Component: Water pollution recognition AI CAM (OVE-C), monitoring and signal directions) • Startup: Sheco, Korea
	1.6.6. Waste • Reducing the site's waste production • Smart waste-treatment system • Separate collection system	• AI trash bin using smart green card • Collection bag and vehicle	**16. Conveyor-type Mass Organic waste treatment device using insects** [Component: Growth environment maintenance software, ICT-based monitoring technology, enabling immediate remote response and management] • Company Foodyworm Agricultural Inc. Korea

(continued)

Table 6.1 (continued)

Life cycle stage of project activities	Application areas	AI technologies (AI framework)	Best practices (Startups)
2. Construction works	2.1. IoT infrastructure construction 2.2. Construction site accident, prevention and safely management 2.3. Construction scheduling software 2.4. Construction costs	• AI physical-ecological-digital integration of infrastructure • AI-generated 3D construction site • Spatial intelligent technology • Construction robot	**17. AI-based convergence service** [Component: Intelligent transportation, autonomous driving, smart construction, digital twin, digital human, etc.] • Startup: METABUILD, Korea **18. AI smart construction management** [Component: On-site DX service to increase Productivity of construction management and works reporting and dashboard management] • Company: AIdagram Inc., Japan **19. Saas NUBI SON AIoT** [Component: YOLO 7-based smart safety algorithm Intelligent chatbot—safety measure notice service] • Company: Simplatform Inc., Korea
			20. AI-powered construction process management solutions [Component: Advanced AI and Deep-learning data analytics, smart -data driven decisions] • Startup: UPSIGHT, Korea

(continued)

Table 6.1 (continued)

Life cycle stage of project activities	Application areas	AI technologies (AI framework)	Best practices (Startups)
3. Operation and maintenance	• Operation center with supercomputing cloud server: Cloud 5G data center • Telecommunication. Services • Monitoring station operation • Optimization: Optimal (hybrid) operation • Maintenance center operation • Smart management: Management tracker • Overtourism management	• AI digital visitor count • Digital security • ChatGPT for nature interpretation • Internal connections (Municipal WiFi broadband, 5G/5.5G/6G) • Disaster (flooding) software • AI bike count • AI rubbish bins • AI missing child trackers • AI irrigation system • AI mist sprayers • AI locker • AI lifestyle: Consumption pattern and behavior • AI controller • AI wireless CCTV camera watch • Communication with WiFi • Robots: Smart controllers • Virtual station technology • Cloud-based management platform • Disaster data reporting	**21. AI phone** [Component: Galaxy AI phone] Company: Samsung, Korea **22. AI chat GPT** [Component: Data source, Big data platform, AI platform, customer service] • Startup: PERSONA AI, Korea **23. AI cloud 5G Data Center** [Component: private 5G network, data collection and transmission, data-based intelligent solutions] • Startup: WIZCORE Private 5G, Korea **24. AI Blockchain-based NFT (Non Fungible Token) technology** Component: Cloud-based Saas Service from advance ticket to admission • Institution and startup: nipa, Rootlab, Korea **25. Secure Today, for a Better Tomorrow: Palo AI to Networks** [Component: Machine Learning (ML)-based next generation firewall (NGFW), threat prevention, advance URL filtering, IoT security, 5G-native security, prisma cloud, etc • Company: i-Cloud and Palo Alto Networks, Chicago, USA **26. Location data-based visitor tendency analysis** Component: Data collection, multidimensional cluster analysis, judgement of visitor tendency • Startup: NUBISON AIOT, Korea

(continued)

Table 6.1 (continued)

Life cycle stage of project activities	Application areas	AI technologies (AI framework)	Best practices (Startups)
4. Monitoring	• MRV • Real time on-site monitoring • Automatic remote monitoring	• AI remote control solutions • Quantum computing structure • Interconnected sensor networks and Internet of Things (IoT) – Remote monitoring meters: Remote sensing techniques – SCAD/ IoT: Supervisory control and data acquisition – Acoustic monitoring sensors – Ecological and environmental sensors – AI water robotics with camera to get information on water pollution – Water level and quality monitoring: Telemonitoring sensors – Soil-moisture monitoring sensors – Vegetation-distribution and health-monitoring sensors – Plant-photosynthesis monitoring sensors – Meteorological sensors. (temperature, humidity, wind, etc.) – Carbon sink and storage monitoring sensors – Noise monitoring sensors – Fine-dust monitoring sensors – Air-quality monitoring sensors – Water-pollution monitoring sensors – Intelligent water sensors • Supercomputing cloud server for monitoring data storage	**27. U$^+$5G** • **Company: LG U$^+$, Korea** **28. The 1st Digital Inspection** [Component: AI-based vision inspection solution, YOLO-based AI model to detect defects, algorithm processing technology] Company: DWORLD, Korea **29. AI-based autonomous driving drone pollution source monitoring system** [Component: Image recognition of people and sources of pollution] • Corporation: Kwater, Korea **30. AI Agritech analytics** [Component: Integrated satellite data analytics, and data from solar-powered IoT (crop and soil) sensor to halt and reverse soil degradation, crop pests and diseases, and to monitor plant photosynthesis, combination of the satellite data and sensor data] • Startup: Agritech Analytics, Kenya **31. Algae boom detection** [Component: Enabling rapid detection and analysis for large-scale algae blooms for prompt response actions, analysis of the causes of algae blooms and provision information for water quality management and eco-system protection] • Startup: Stellarvision, Korea

(continued)

Table 6.1 (continued)

Life cycle stage of project activities	Application areas	AI technologies (AI framework)	Best practices (Startups)
		• Automatic real time digital meters • Communication network with information terminal (internet with 5G/5.5G/6G and WiFi in and around tourism site)	
5. AI capacity building, education and citizens participation in tourism and wetland site management	• Metaverse experiencing space • AI for creative smart life: Smart living • AI appliances • Demand side management • Communication by digital means • Promoting the participation of citizens in tourism activities • Minimizing the ecological footprint of visitors • Community garden practices • Safe eco-trails • Outsourcing digital talents • Strengthening partnerships	• Personalized AI • Metaverse technology • Video analytics and citizen apps • AI copyright proposals • Smart community farming technology • Integrated CCTV camera watch	**32. AI Blockchain technology-based services** [Component: Building an integrated advertising program based on Blockchain] • Startup: Rootlab, Korea **33. AI agriculture robot** • Startup: MiFood, Korea **34. AI and IoT communication Program** [Component: Multilingual communication program] • Company: Kotozna, inc., Japan **35. Education technology** [Component: Combination of AI technology and talent management to build digital talent marketplace platform • Startup: Umurava, Rwanda

References

ChatGPT (2025) website
Korean Air (2025a) Pixels and Poems: seattle, pp 22–25
Korean Air (2025b) The poet of algorithms: Refik Anadol, pp 98–101
Schwab K (2016) The Fourth Industrial Revolution, Penguin Random House, UK, p 1
Snowflake (2025) Snowflake World Tour, COEX, Seoul, Korea
UbiQuoss (2025) AI-based UNIS PRO: integrated IoT networking system of buildings (apartments, houses, etc.) connected with AI agent
Zheng U (2018) Urban computing, The MIT Press, Cambridge, pp 302, 313, 469

Chapter 7
AI Urban Innovation Ecosystem Strategies for City and Community-Based AI Transformation

Kwi-Gon Kim

Abstract We need AI transformation for new forms of urbanism, and it serves several purposes. One is to ensure the connection of big data and AI platforms for the data-driven AI cities described in the previous chapters. Another key purpose of AI transformation is the evaluation of optimization solutions in relation to sustainability. AI transformation strategies are required for creative disruption technologies to ensure compliance with the spirit and intentions of UN SDGs, criteria and indicators for sustainability. Siebel (Digital transformation: survive and thrive in an era of mass extinction. Rosetta Books, New York, p 11, 2019) states that:

> I find it more valuable to describe digital transformation through examples. This is partly because we are in the midst of massive disruption and constant change. The scope of digital transformation and its implications are still evolving, and its impacts are still being understood. Each iteration—whether across companies or industries or even within a single organization—will bring new insights and layers to our understanding of digital transformation.

This chapter addresses the strategic principles of AI transformation for climate solutions, plans for profiting from AI creative disruption technology, new tactical project proposals, and AI infrastructure policies.

Keywords AI transformation strategy · Strategic principles · Creative disruption technology platform · Fairness · Knowledge gap · Feedback loop · Just transformation · AI analyst · Access

---

K.-G. Kim (✉)
Seoul National University, Seoul, South Korea
e-mail: kwigon@snu.ac.kr

K.-G. Kim (ed.), *AI Transformation of Cities for Climate Solutions*,
https://doi.org/10.1007/978-3-032-21940-4_7

7.1 Key Strategic Principles of AI Transformation for Climate Solutions

Implementing AI to address climate solutions involves much more than simply installing new software or building new AI infrastructure. An AI transformation requires a fundamental shift in how a city is planned and designed, constructed, operates, makes decisions and delivers additional values.

To succeed, cities need to follow a set of key principles that balance AI technical capability and a human-centric approach.

Key strategic principles of AI transformation to tackle climate challenges are listed as follows:

- Standard: To create an AI transformation standard to help cities understand how their activities can respond to climate change, including a public register of algorithms
- Knowledge gaps: To identify knowledge gaps in the AI transformation from traditional to innovative practices
- Feedback loop: To use a feedback loop in AI transformation within city climate planning
- Justice: To ensure just AI transformation in relation to AI ethics, including uneven access to data and the digital divide
- Coalition: To organize a coalition for sustainable AI networking for climate solutions
- Multi-AI agents: To develop a list of multi-AI agents to tackle climate challenges
- Data quality and readiness: To identify data needs and maintain clear, reliable and unbiased data to guide AI transformation strategies
- AI workflows: To generate the flow of AI tools associated with AI transformation
- Compliance: To provide the degree of compliance of AI transformation required for selected sustainable development goals
- Interdisciplinary team: To use a systematic, interdisciplinary and organized approach to AI transformation for climate solutions
- Real and virtual world: To predict and optimize anticipated climate impacts through AI applications of both mathematical and visual presentation techniques
- Fairness and inclusivity: To actively test for and mitigate biases in data and algorithms
- Replicability, transparency and accountability: To avoid different results which can occur depending on the AI analyst
- Human-centric AI: To demonstrate the potential of combining human insights with AI by involving citizens in training a machine learning algorithm.

7.2 List of AI Transformation Strategies

AI drives creativity and provides creative disruption technologies for city climate planning. The following are emerging AI transformation strategies that improve current practices with AI creative disruption technology platforms:

- Strategy 1: Optimization of future urban sustainability based on AI demand prediction, using sustainability goals, criteria and indicators.
- Strategy 2: Switching greenhouse gas (GHG) emissions accounting systems from production-based systems to needs/consumption-based personalized AI systems. Consumption/needs-based GHG accounting can switch from annual to real-time or hourly accounting, with AI creative disruption technology. The switch can be achieved by using personalized AI GHG tracking AI apps and edge AI (local AI: mobile phones, etc.) to analyze user behaviour and purchases, as well as consumption of goods and services to estimate carbon footprints and suggest personalized emissions strategies.
- Strategy 3: Adoption of a multi-AI agent system (MAS) as a tool for integrating environmental, social and economic considerations into the optimization process. An MAS is a collection of autonomous, interacting computational entities called agents that work together to solve complex problems. These agents can collaborate, coordinate or compete to achieve individual or collective goals, using distributed control instead of a single central point. Therefore, an MAS is more holistic than a multi-tiered assesssment based on the map overlay method. Multi-AI agents use a more holistic framework that contains different agent (player) types such as a wetland habitat agent, a water or hydrology agent, a climate agent, a public participation agent and an AI eco-cultural tourism agent (as shown in Fig. 4.1).
- Strategy 4: Tokenization of AI climate tech credits. AI can be leveraged for predictive baseline modeling and automated compliance auditing. The promise of AI-enabled carbon credit markets, including the tokenization of credits using distributed ledger technology, can lower barriers to entry and ensure that urban climate projects—such as energy efficiency retrofits, urban forestry and community solar—can be a source of much-needed climate finance for jurisdictions and communities. Also, AI creative disruption technology can be used to tokenize the credits that are generated by application of AI climate tech in climate solutions.
- Strategy 5: A backcasting approach based on normative goals vs an AI demand forecasting approach based on real-time data. The backcasting approach is a strategic method that begins by defining a desirable future state and then works backward to identify policies, strategies and actions needed to achieve zero-carbon emissions. On the contrary, AI demand forecasting uses artificial intelligence, specifically machine learning and predictive analytics, to analyze historical and real-time data to predict future demand for products and services. This method is more accurate than traditional forecasting because it incorporates a wider range of factors, such as sales data, consumer behavior, market trends and external signals like weather and promotions, and continuously learns from new information to

refine its predictions. The goal is to help businesses optimize inventory, improve supply chain management, reduce waste and make better decisions regarding production, pricing and strategic planning (Google 2025).

- Strategy 6: AI startups for business 5.0. AI startups are characterized by their unique approach to integrating artificial intelligence into their business models and operations. These characteristics include innovative use of AI. They stand out by creatively applying AI to develop unique solutions or disrupt markets.
- Strategy 7: Local participation. Civic participation based on AI citizen science, which focuses on engagement with the public, can yield benefits for governments and citizens alike.
- Strategy 7: Sustainable supply chain. Using AI for life cycle analyses (LCA) to measure the carbon footprint of consumer goods and services, and guide transitions to more sustainable and low-carbon products.
- Strategy 8: Innovation in sustainable site design, operation and maintenance. Designating areas as protected habitat and open space, and developing AI-driven sustainable site managemenrt plan (U.S. Green Building Council, 800, 795, 1747).

7.3 Strategic Project Proposals to Transform Cities with Next Generation AI

This section proposes some application areas for creative disruption technology to optimize urban systems.

AI brings tangible improvements to urban living by enabling cities to respond to challenges with speed, precision and foresight. These benefits span across mobility, sustainability, safety and infrastructure resilience (Table 7.1).

The project proposals are built upon concepts of Industry 5.0 or Business 5.0 which takes a human-centered approach, emphasizing the collaboration between humans and machines to achieve unprecedented levels of efficiency, productivity, sustainability and customization.

Table 7.1 Benefits of AI disruption technologies

Benefit	Description
Improved mobility	AI-driven systems reduce traffic congestion and improve transit schedules
Enhanced public safety	AI helps identify risks and deploy emergency services faster
Energy optimization	Smart grids use AI to match supply and demand efficiently
Waste reduction	AI-powered logistics streamline collection and recycling efforts
Predictive maintenance	Infrastructure health is monitored in real time to prevent costly failures
Inclusive planning	AI helps design more equitable, accessible urban spaces

Industry 5.0 represents the next phase of the industrial revolution, where advanced technologies, such as Internet of Things, AI, robotics and data analytics to enable real-time data collection, analysis and decision making, could converge to create intelligent, collaborative planning environments. In embracing this transformative concept, the following new strategic projects are proposed as examples:

- AI-enhanced green life cycle for food consumption at restaurants
- AI-enhanced building upcycling in high-tech architecture
- Ai-enhanced repurposing of scrap fabric waste into new creations
- AI-boundless urban parks brimming with solar energy
- AI-enhanced water-sector infrastructure
- AI-enhanced resilient and sustainable infrastructure systems for a changing climate
- AI-enhanced smart life style (in fashion etc.).

7.4 AI Infra Strategies

Many AI chips with diverse functions are embedded in home appliances, smart phones and various industrial sectors to benefit humanity. Thus, AI infrastructures are needed to combine planning tools and AI agent tools, including AI data centers with cooling solutions for implementation of the strategic projects proposed above. AI infra strategies for facilitating various planning portfolios should be developed for AI lead.

References

Google (2025) website

Siebel TM (2019) Digital transformation: survive and thrive in an era of mass extinction. Rosetta Books, New York, p 11

U.S. Green Building Council (undated) Building and LEED: Green Building and LEED Core Concepts Guide, p 56

Chapter 8
Action Plan for Implementation of AI Transformation in Cities

Kwi-Gon Kim

Abstract It is time to ride the AI wave from curiosity to action! The UK National AI Strategy (2021) and UK AI Opportunities Action Plan are roadmaps for the government to harness AI in order to drive growth, boost productivity and provide tangible benefits for UK citizens. The basic purpose of these documents is to guide the development, adoption and regulation of AI across the country. The strategy represents a comprehensive, 10-year vision for AI as a 'science and AI superpower'. This chapter discusses how cities can leverage new technology to deliver the real value of AI in transforming the way we think, work and live in cities. The chapter discusses how cities can use AI to predict climate change and consumer behavior in tracking consumption-based GHG emissions and to optimize climate action in the entire life cycle of climate-responsive urban planning and development. It also explores how to promote the utilization of AI in the practical delivery of public and private services, including AI infrastructure development.

Keywords AI climate action plan · Moral questions · Bias · Datasets · Sovereign AI · AI regulations · GPU · TPU · AI education · Standardization · AI leadership

8.1 AI Urban Climate Ethics: Moral Questions About Machine Learning Without Permission

Concerns are growing about biases in AI systems (Fig. 8.1). Ethical issues about the use of AI in climate change include its own significant carbon footprint, the perpetuation of inequality through biased data and uneven access, and the potential for its misuse in ways that undermine sustainable development goals. Other concerns involve transparency and accountability, the environmental impact of AI-driven consumption, the displacement of jobs and strategic value added through sustainable operations. All of these complicate AI's role as a climate solution.

K.-G. Kim (✉)
Seoul National University, Seoul, South Korea
e-mail: kwigon@snu.ac.kr

K.-G. Kim (ed.), *AI Transformation of Cities for Climate Solutions*,
https://doi.org/10.1007/978-3-032-21940-4_8

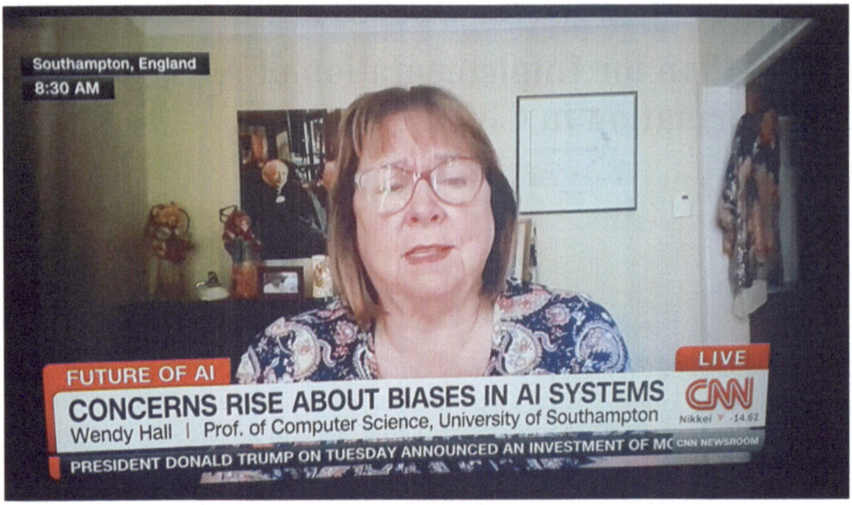

Fig. 8.1 CNN's interview with Professor Wendy Hall about the moral issues of AI. *Source* CNN, 16 July 2025

8.2 Climate Data Center and Sovereign AI

Climate data can be stored and accessed using large-scale, cloud-based servers like the *Copernicus Climate Data Store (CDS)* and national archives like *NOAA's Climate Data Online*, or through private/on-premises infrastructure using technologies like *Ceph storage* for scalable, distributed storage. These systems provide access to a wide range of datasets, including historical data, observations, and climate model projections, along with tools for analysis and processing (Figs. 8.2 and 8.3).

However, most database services are provided for weather data stored at the national and global level. AI climate data servers in AI data centers should be provided for downscaled climate data at the local and regional level.

Furthermore, there is an issue of sovereign AI (Box 8.1 & Fig. 8.4). Sovereign AI is AI that is produced using a nation's own infrastructure, data, workforce and business networks. It is built independent of foreign influence and is specifically designed to reflect the nation's identity and autonomy. Choosing a sovereign data center with climate servers is vital for safeguarding and future-proofing AI investment and protecting this valuable asset.

Box 8.1 Sovereign AI for the era of enterprise AI. *Source* Macquarie Data CENTRES, Sovereign AI: Why it matters, eBook, 2025

Fig. 8.2 Connection of GPU, HBM and HBF. *Source* Chosun Daily News, 12 November 2025

Fig. 8.3 Data center embedded with Google TPU (Tensor Processing Unit). *Source* Chosun Daily News, 28 November 2025

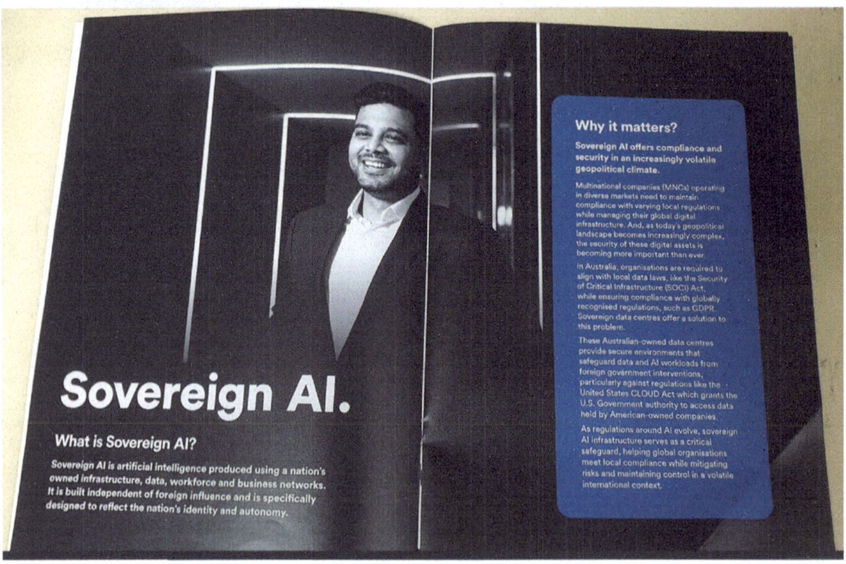

Fig. 8.4 Sovereign AI. *Source* Macquarie Data CENTRES, Sovereign AI: Why it matters, eBook, 2025

8.2.1 Other AI Implementation Issues

Other implementation issues for AI are listed as follows:

- AI regulations
- Connecting the AI climate action plan at all levels to bridge local climate action with national climate action plans, and with the implementation of global SDGs and the Paris Agreement, including natural climate solutions
- AI planning education: Incorporating AI into planning and environmental education focusing on Business 5.0 or Industry 5.0
- Practical applications: Education, upskilling and reskilling for the AI green transition
- Securing AI consumer data rights
- Institutional innovation: Establishment of new organizations
- Procurement of semiconductors including GPU and Tensor Processing Unit (TPU) which are

  core hardware for AI operation in order to build AI industry infrastructure
- Support for research and development (R & D) and human resources associated with AI technology
- Integrated implementation mechanism: Governance for realization of AI democratization based on the concept of citizen science and data science
- AI criteria and standardization and copyright for the collaborative work of humans and AI
- AI performance standardization by ISO
- AI leadership: AI-driven cities will rapidly advance in AI with the development of AI infrastructure.
- AI for city competitiveness: AI can enhance the brand value of cities
- The establishment of International AI Cooperation Organization: Silicon valley launches 'Pro-AI PACs' (25 August 2025).

References

CNN News (2025) Future of AI: Concerns rise about biases in AI systems, 16 July
Macquarie Data Centres (2025) Sovereign AI: Why it matters, eBook
UK Government (2021) UK National AI Strategy

Part III
Status and Future Trends in AI

Chapter 9
Current AI Research on Cities for AI Innovation

Kwi-Gon Kim

Abstract AI research on cities is rapidly expanding, encompassing smart cities, ubiquitous cities, virtual cities, AI cities and 'urban informatics'. The effectiveness, efficiency and practicality of development planning have increasingly been challenged by the complex nexus between physical, spatial, ecological, social, economic and climate change problems associated with sustainability, conflicting interests of stakeholders and citizen activism. This chapter combines in-depth analysis of current AI research, existing case studies and viewpoints from those on the frontline of AI research and practice.

Keywords Virtual city · Startup · Explainable AI · AI regulation · City scale · Coalition for sustainable AI · AI ethics · Artificial general intelligence · Forecasting demand · Innovation hub · Misinformation · Satellite imagery · Business process optimization

9.1 Comparison of AI Status at Present and Trends for the Future

Figure 9.1 shows the status of AI at present and trends for the future (Sin 2025). AI transformation of cities for climate solutions is relatively new and as such, represents a developing technology.

An area of future importance in AI agents is associated with governance. More efforts should be made to develop an AI agent governance framework of policies, processes and technical controls to provide both command and transparency over AI systems throughout their existence. A proper governance layer must include the entire process of data preparation, model selection, tool integration and evaluation, and monitoring.

K.-G. Kim (✉)
Seoul National University, Seoul, South Korea
e-mail: kwigon@snu.ac.kr

K.-G. Kim (ed.), *AI Transformation of Cities for Climate Solutions*,
https://doi.org/10.1007/978-3-032-21940-4_9

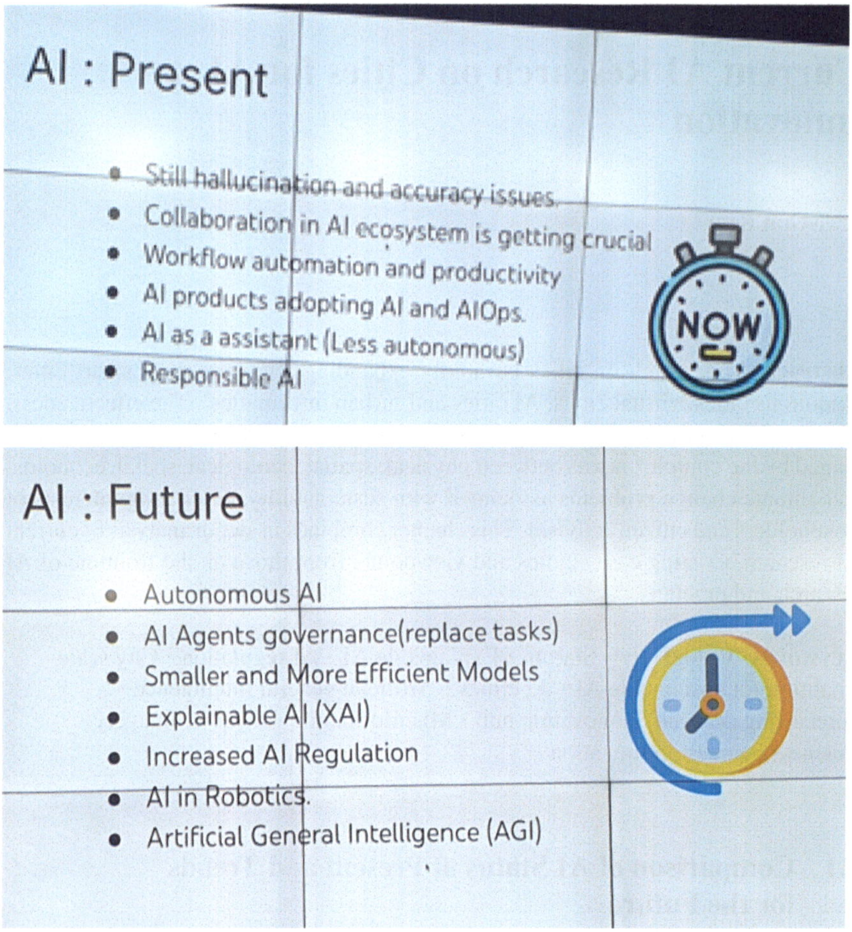

Fig. 9.1 Comparison of AI status at present and in the future. *Source* Sin (2025) Generative AI Trends and AI Agent, paper presented at Google Startup Campus, Seoul, Korea, 17 June, pp. 44–45

9.2 How Cities Use AI Technologies

According to the Australian Trade and Investment Commission (2025), small and medium-sized companies are likely to adopt AI for marketing automation, predictive analytics, fraud detection, supply-chain optimization, sales forecasting, data entry and document processing, customer support and chatbots, AI assistants and employee recruitment ranging from 11 to 22% (Fig. 9.2).

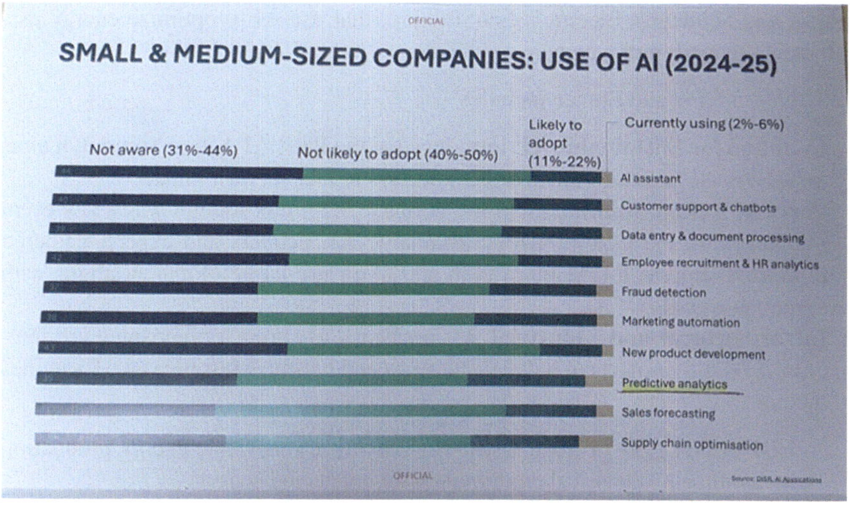

Fig. 9.2 Small and medium-sized companies use of AI. *Source* Australian Trade and Investment Commission, Australia's data center landscape: Current and future potential for AI and clean energy, 2025

9.3 AI-Driven Cities Creating a New Form of Climate Capital

Cities are reclaiming space and infrastructure for people and improving urban life. AI case studies associated with climate change can be reviewed at different levels:

- AI cases at the city scale
- AI cases at the neighborhood scale: Real-time data on various aspects, such as the flow of people, noise levels, temperature, CO_2 concentration levels and air quality in each neighborhood to power AI-accelerated digital twins
- AI cases at the block scale
- AI cases at the building scale.

This section reviews seven cases of AI applications at the city scale, with a focus on climate solutions through their websites.

9.3.1 AI in Paris, France

Paris (2025) is actively engaged in using AI to combat climate change, most notably through the Coalition for Sustainable AI launched at the Paris AI Action Summit, which promotes using AI for environmental benefits while addressing its own footprint. France also has domestic initiatives, such as EDF's smart grid project and Schneider Electric's EcoStruxure, that use AI to optimize energy use, reduce waste, and improve energy efficiency to fight global warming.

Initiatives in Paris and France include:

- **Coalition for Sustainable AI**: Introduced at the Paris AI Action Summit, it aims to harness AI for environmental benefits and reduce its own impact.
- **Paris AI Action Summit**: Held in February 2025, this summit was a key event for launching the Coalition for Sustainable AI. Leaders and experts gathered to discuss support for AI innovation and to ensure its development aligns with sustainable goals.
- **Infrastructural and industrial AI applications**: French companies are integrating AI into projects to reduce emissions and optimize resource use. Examples include:
  - **EDF smart grid project**: Uses AI to optimize renewable energy production by monitoring electrical equipment with connected sensors.
  - **Schneider Electric's EcoStruxure**: Employs sensors and machine learning to monitor and adapt energy consumption in buildings and industries, thereby improving energy efficiency.
  - **SUEZ's eco-industrial park**: Utilizes AI to optimize waste treatment and maximize material and energy recovery from waste.
- **Sustainability and ethics**: The initiatives above all aim to place sustainable AI considerations and AI security and ethics at the forefront of global discussions.

9.3.2 AI in Berlin, Germany

Berlin (2025) is a prominent hub for artificial intelligence (AI) development and application in Europe. Key aspects of AI use include:

- **Strong AI ecosystem**: Numerous companies, research institutions and initiatives drive innovation in various sectors
- **Diverse AI applications**: AI is widely used across a range of applications including healthcare, mobility and business process optimization
- **Focus on innovation**: Berlin is a leader in AI research and development, with a focus on areas like explainable AI (XAI) and open source AI.
- **Platform for collaboration**: Berlin is a dynamic center for AI innovation, with a strong ecosystem, diverse applications and ambitious plans for future growth.

9.3.3 AI in London, UK

London's (2025) AI sector contributes to various fields, including:

- **Self-driving vehicles**: Through companies like Wayve

- **Video generation**: Via AI avatars using Synthesia
- **Data intelligence**: For finance, insurance and telecommunications sectors, as seen with Quantexa
- **Cybersecurity:** With firms like Quantexa developing AI to secure data and detect illegal activities.

9.3.4 AI in New York City, USA

Key applications include:

- **Applied AI**
- **Healthcare**
- **Climate change**: AI technologies are being developed to address and create solutions for climate change
- **Media**: The *New York Times* uses AI for automated voice technology to make articles accessible and to assist with translations.

In New York City (2025), AI is being used for climate change adaptation and mitigation by forecasting and managing the electrical grid, improving building energy efficiency, mapping environmental risks like flooding and heat stress and developing more cost-effective infrastructure for flood defenses. Initiatives like.

New York Institute of Technology's research and the Empire AI initiative are leveraging AI to help city officials make data-informed decisions to prepare for and address urban environmental challenges.

AI climate solutions include:

- **Forecasting demand**: AI algorithms, partnering with utilities like Con Edison, use vast amounts of weather data to forecast electrical load demand, which is crucial for managing the grid during New York's increasingly hot summers
- **Optimizing EV charging**: AI can manage when electric vehicles charge to avoid peak loads on the grid, helping to reduce strain and allow for more renewable energy integration
- **Predicting outages**: Machine learning models help utilities predict power outages by analyzing weather forecasts and past storm impacts, allowing for a faster response and quicker restoration of power
- **Urban planning and environmental risk**:
  - *Mapping environmental risks*: Researchers use AI to create data-informed maps of flash flooding, building energy efficiency, heat stress and fallen trees.
  - *Data-driven decision making*: AI-based tools empower city officials to make better decisions for neighborhood-level energy efficiency and resilience to environmental issues
  - *Flood defense design*: AI-powered methods, such as reinforcement learning, can design dynamic seawalls that adjust their height over time based on projected sea-level rise, making defenses more efficient and cost-effective than traditional static designs.

9.3.5 AI in Barcelona (2025), Spain

Key aspects of AI in Barcelona include:

- **Innovation hub**
- **Industry and startups**: The city hosts many AI consulting firms and startups in sectors like health tech, legal SaaS, and agricultural AI, with international players such as NTT Data, Microsoft, Sony AI, and Foundever establishing deep tech centers.
- **Smart city and tourism**: The city is a pioneer in smart city solutions, applying AI for traffic management and urban planning. AI also enhances the tourism sector through personalized experiences and chatbots.
- **Ethical AI Focus**: Barcelona is committed to the ethical development of AI, with a municipal strategy focused on transparency, data rights and the use of algorithms in public services. This includes the creation of a public register of algorithms and an international AI observatory.
- **Events and education**: The city is a hotspot for AI and tech events, such as the Mobile World Congress (MWC) and the Big Data & AI Congress. Educational initiatives like the community-driven 'AI Saturdays' aim to make high-quality AI education accessible to everyone.
- **AI climate modeling and urban planning**: Urban living arm. Here are examples of how the city has been helping to shape the future of urban living:
  - *Digital Twins of the City*: Barcelona's virtual replica, the first of its kind in Europe, marks an urban planning revolution. It allows simulations of various scenarios, modeling the impact of potential changes, like the installation of a new metro line or the effects of a heatwave on vulnerable populations. These digital twins are further enhanced by AI, powered by the BSC-CNS, enabling real-time tracking of air and noise pollution, distribution of goods, traffic management and building energy use to better understand an incredibly complex urban system.
  - *Sentilo*: An open-source software that utilizes thousands of sensors installed in public spaces throughout the city to capture real-time data on various aspects, such as the flow of people, noise levels, temperature and air quality in each neighborhood. This data provides valuable insights into the city's dynamics, supplementing the information used to power AI-accelerated digital twins, and enabling more effective urban planning and resource management.
  - *Mercè*: This citizen science experiment conducted from May to November 2020, demonstrates the potential of combining human insights with AI. By involving citizens in training a machine learning algorithm to analyze evaluations of over 3,000 streets, Mercè helps identify urban design patterns that contribute to a more livable city, such as the presence of trees, wide sidewalks or urban furniture.
  - *Evolving digital awareness in urban environments*: As urban environments grow more complex, the demand for intelligent solutions is greater than

ever. Urban AI is an emerging concept that is an extension of the Smart City paradigm, helping cities navigate the challenges of the twenty-first century, and Barcelona is at the forefront of this revolution.
- *GOUAI's Atlas of Urban AI*: On July 20th, 2022, Barcelona, along with Amsterdam and London, presented the Global Observatory of Urban AI (GOUAI) at the Barcelona Urban Innovation Center. The GOUAI's Atlas of Urban AI serves as a global platform, providing a visual representation of how cities worldwide are using AI to address a wide range of challenges.

9.3.6 AI in Toronto, Canada

In Toronto (2025), AI is being used for climate solutions in various approaches, including the University of Toronto's Sinton Lab's self-driving lab for developing new materials to capture carbon, Evergreen's 'AI for the Resilient City' program predicting urban heat island effects and aiding climate planning, and Manifest Climate's use of AI for sustainability intelligence to help businesses transition to a low-carbon future.

Research and development (R & D) include:

- **Sinton Lab at the University of Toronto**: This self-driving lab uses AI to accelerate the discovery of new materials and molecules to convert captured carbon dioxide into valuable products like ethanol and ethylene
- **CIFAR**: This institute is advancing AI to help model the climate and speed up research in areas like biodiversity, renewable energy, and crop yields.

Additional R & D for urban planning and adaptation include:

- **AI for the Resilient City Program**: Developed by the Toronto and Region Conservation Authority (TRCA) and Evergreen, this program uses AI and satellite imagery to create custom scenarios for urban heat islands (UHIs). It helps communities understand and plan for extreme heat events by analyzing factors like built environments and demographic data, according to www.evergreen.ca.
- **Manifest Climate**: This Toronto-based company provides AI-powered sustainability intelligence to help businesses navigate the transition to a low-carbon economy by managing and reporting on their climate impact.
- **Green Light**: While not Toronto-specific, this AI-based application can optimize traffic light scheduling to reduce stop-and-go traffic, which, in turn, lowers fuel consumption and emissions.

Broader climate applications include:

- **Climate modeling**: AI algorithms are being developed to speed up complex climate models used by the Canadian government, making long-term climate predictions more accessible.

- **Energy efficiency**: The Vector Institute is collaborating on open-source AI solutions to help large organizations reduce their energy footprint and overall emissions.

9.3.7 AI in Montreal, Canada

Montreal (2025) leverages AI to combat climate change through initiatives like the Mila-developed website "This Climate Does Not Exist," which uses realistic imagery to raise public awareness of climate impacts. The Montreal Insectarium uses AI to monitor insect populations for the Antenna project, helping to address biodiversity loss due to climate change. Additionally, Montreal-based BrainBox AI uses AI to optimize building energy consumption and reduce emissions, showcasing the city's leadership in applying AI for environmental solutions.

Examples of AI for climate action in Montreal include:

- **Raising public awareness**:
  - **This Climate Does Not Exist**: Founded by Mila's Yoshua Bengio, this website uses AI to generate realistic images of what various locations would look like under different climate change scenarios, such as flooding, wildfires or smog. The goal is to make climate change impacts more personal and to spur empathy and action.
- **Protecting biodiversity**:
  - **Antenna Project**: Led by the Montreal Insectarium, this project uses AI and high-resolution imagery to track declining insect populations and identify new species. The initiative aims to document biodiversity loss caused by climate change, pollution and habitat destruction.
- **Reducing energy consumption in buildings**:
  - **BrainBox AI**: This Montreal-based company equips buildings with AI systems to predict and adjust heating and cooling needs, significantly reducing wasted energy and emissions. The technology can improve energy efficiency and lower utility costs.
- **Institutions driving AI for climate action**:
  - **Mila Quebec Artificial Intelligence Institute**: A world leader in AI research, Mila is involved in various human-centric AI projects. They developed the 'This Climate Does Not Exist' website and also run the AI for Climate Studio, which develops and distributes AI tools to help climate actors.
  - Montreal Insectarium: A part of the Montreal Space for Life, it leads the Antenna project, applying AI to understand and address the environmental crisis.

9.3.8 AI in Vancouver (2025), Canada

Key areas of focus include:

- **Climate data analysis**: Using AI to process large datasets to better understand climate patterns and ocean processes
- **Sustainable infrastructure**: Applying AI to design and manage infrastructure that is more resilient to climate impacts and contributes to zero-emission mobility
- **Resource optimization**: Employing AI to improve the efficiency of industrial and urban systems, reducing energy consumption and emissions
- **Accelerating research**: Developing AI tools that expedite climate simulations and environmental research, providing faster insights for policymakers and scientists.

9.3.9 Washington D.C. (2025), USA

Applications of AI in climate change efforts include:

- **Sustainable energy**: Optimizing power-grid efficiency and facilitating the integration of renewable energy sources
- **Climate research and decision making**: Aiding climate-related research, improving forecasting and supporting more effective climate policy
- **Biodiversity and nature conservation**: Developing AI tools to monitor and protect biodiversity and nature
- **Combating misinformation**: AI-powered tools, like the CARDS project, are being developed to detect and debunk climate change denial and misinformation
- **Sustainable supply chains**: Using AI for life cycle analyses (LCA) to measure the carbon footprint of consumer goods and guide transitions to more sustainable products.

Research and development for challenges and considerations include:

- **AI's environmental impact**: Acknowledging that the significant energy consumption of AI data centers presents challenges and necessitates strategies for mitigating these environmental effects
- **Responsible implementation**: Ensuring that AI solutions align with principles of equity and climate justice is a key focus
- **Data and infrastructure**: Developing and sharing the necessary datasets, tools and research infrastructure crucial for the successful deployment of AI for climate action.

9.4 Comparison of City Level AI Systems

Table 9.1 presents a summary comparing the use of AI driving new approaches to urban planning and management and AI for climate solutions at the city level in a selection of the cities discussed in this chapter.

Many uses of AI at the city level have involved climate monitoring and sectoral climate solutions. However, there are essentially no items in Table 9.1 that are oriented to the sustainable cost–benefit analysis, sustainable goal achievement and climate impact assessment to ensure the integrated use of AI in planning and decision making.

Table 9.1 Comparison of AI systems in selected cities

City	Uses of AI in urban planning	AI actions for climate change at city level
Paris	• To support AI innovation and ensure that its development aligns with sustainable goals • To optimize smart-grid projects • To optimize energy use • To optimize renewable energy production by monitoring electrical equipment with connected sensors • To employ sensors and machine learning to monitor and adapt energy consumption in buildings and industries, thereby improving energy efficiency • To optimize waste treatment and maximize material and energy recovery from waste	• To analyze data for environmental monitoring, such as mapping CO_2 and methane emissions
New York City	• To use vast amounts of weather data to forecast electric load demand, which is critical for managing the grid during New York's increasingly hot summers • To optimize electric vehicles (EV) charging to avoid peak loads on the grid, helping to reduce strain and allow for more renewable energy integration • To use machine learning models to help utilities predict outages by analyzing weather forecasts and past storm impacts, allowing for a faster response and quick restoration of power	• To develop AI technologies to address and create solutions for climate change • To use AI to create data-informed maps of flash flooding, building energy efficiency, heat stress and fallen trees • To empower city officials through AI-based tools to make better decisions for neighborhood-level energy efficiency and resilience to environmental issues • To design (with AI-powered methods such as reinforcement learning) dynamic seawalls that adjust their height over time based on projected sea-level rise, making defenses more efficient and cost-effective than traditional static designs

(continued)

Table 9.1 (continued)

City	Uses of AI in urban planning	AI actions for climate change at city level
Toronto	Not available	• To develop new materials to capture carbon • To predict effects of urban heat islands • To aid climate planning • To use AI for sustainability intelligence to help businesses transition to a low-carbon future • To accelerate the discovery of new materials and molecules to convert captured CO_2 into valuable products like ethanol and ethylene • To help AI model the climate and speed up research in areas like biodiversity, renewable energy and crop yields • To use AI and satellite imagery to create custom scenarios for urban heat islands (UHIs) • To optimize traffic light scheduling to reduce stop-and-go traffic, which in turn lowers fuel consumption and emissions • To use AI algorithms to speed up complex climate models, making long-term climate prediction more accessible • To collaborate on open-source AI solutions to help large organizations reduce their energy footprint and overall emissions
Montreal	• To reduce energy consumption in buildings	• To combat climate change through initiative like the Mila-developed website

References

Australian Trade and Investment Commission (2025) Australia's data center landscape: Current and future potential for AI and clean energy

Barcelona (2025) website

Berlin (2025) website

London (2025) website

Montreal (2025) website

New York (2025) website

Paris (2025) website

Sin H S (2025) Generative AI Trends and AI Agent, paper presented at Google Startup Campus, Seoul, Korea, 17 June, pp. 44–45

Toronto (2025) website

Vancouver (2025) website

Washington D.C. (2025) website

Part IV
Practices Shaping AI Technologies for Cities

Chapter 10
AI Creative Disruption Technologies: Evidence from Practices

Kwi-Gon Kim

Abstract AI is an emerging force in the creation of disruption technologies. AI servers (such as Google Cloud or Alibaba Cloud) are being integrated into all aspects of city operations including climate change, public safety, planning, transportation and citizen services. This chapter shows how AI technology can be incorporated into cities and used not only for public goods and services, but also for personalized AI. A combination of AI technologies can be utilized to make cities smart and intelligent. Many of the featured technologies are completely new and the chapter provides specific examples to help readers understand the opportunities available and how technology has been, or could be, applied in urban settings. It discusses AI unit technologies, AI system technologies and AI hybrid technologies. All of which are commercially available.

Keywords AI creative disruption technology · AI hybrid technology · AI frontiers · AI-powered disruption · Search engine optimization · Personalized service AI · Virtual software · Climate risk management software

10.1 AI Technologies and Carbon Footprint Reduction

The combined technologies of climate change measures and AI for smarter and greener efficiency-driven solutions can significantly contribute to the reduction and mitigation of the carbon footprint in cities. This section identifies and explores the AI technological opportunities available in urban settings.

K.-G. Kim (✉)
Seoul National University, Seoul, South Korea
e-mail: kwigon@snu.ac.kr

K.-G. Kim (ed.), *AI Transformation of Cities for Climate Solutions*,
https://doi.org/10.1007/978-3-032-21940-4_10

10.1.1 AI Frontiers by Category of Climate Change Measures

There are the three options available to address climate change: adaptation, mitigation and geoengineering (Dessler 2012). Table 10.1 shows the AI technology applications for each of these climate change measure options.

10.1.2 How to Use Search Engines for Climate Action AI Technologies

Generative AI like ChatGPT can help locate and implement more of the climate action measures listed in Table 10.1. Search engines for climate action measures could also change, as a result of the ubiquitous adoption of ChatGPT and other generative AI models.

If we want to combat climate change with mitigation, adaptation and cross-cutting climate action measures, a machine could search, sort and rate candidate measures or technologies. Some measures that could be most qualified might still not be considered for the position for want of the right keywords. Poorly written measure descriptions and requirements may result in a collection of inadequately performing keywords that researchers and consumers might not think to use.

Yet, keywords have persisted because they are generally useful. Beyond sorting many measures quickly, keywords help researchers with important tasks, ranging from research and fact-checking to better-targeted climate action programs.

Search engine optimization (SEO), a process that largely depends on strategic use of keywords to improve the quality and quantity of unpaid traffic to a website or web page, can be utilized. Writing great prompts is also critical to getting great results because ChatGPT works by predicting the words that will follow the words you used in your prompt.

10.2 Personalized Service AI for Consumption/ Needs-Based Greenhouse Gas Accounting

Personalized service AI involves an AI infrastructure with sensors and AI software to enable the service to adapt to changing conditions. The service can improve smart living and consumer behaviors, and consumption/needs-based greenhouse gas (GHG) accounting for climate action.

Examples of the system include:

(1) An AI personalized service for consumption/needs-based GHG accounting for climate action: A citizen science approach for text and image information generation towards customization of AI.

Table 10.1 Summary of AI technologies and carbon footprint reduction

Categories of climate change measures	Methods and techniques	AI technology applications: AI frontiers
Mitigation	• Simulation of population growrh • Urban planning – Transit-oriented development – Infill-densification – Urban blue-green infrastructure Land use – Long-term forecasting – Land-use patterns – Long-term consequences of zoning decisions Urban form – Compact city – 15-min city Energy – Demand control technology – Integrated renewable energy – Electricity grid systems • Buildings – Integrated design process (IDP) into urban planning – Optimizing energy efficiency: District heating/cooling • Waste – Improving waste management – Alternate waste treatments – Waste to energy processes • Water – Water-quality maintenance – Reducing water consumption • Transportation – Enabling next generation mobility – Low-carbon mobility options – Travel demand management – Electric car with autopilots • Lifestyle – Behavior and lifestyle changes – Personal carbon allowance – Municipal clean energy procurement	• AI-powered disruption • AI agents for data processing and model execution for forecasting population growth – Predictive analytics (time series analysis) • AI water and green space grid system • AI innovation to ensure that its development aligns with sustainable goals • AI sustainable goal achievement matrix model • AI land allocation model – AI EIA and climate impact assessment • AI and machine learning (deep learning) for predictive energy-demand forecasting and optimization algorithms • Optimization of smart grid projects and energy use • Optimization of renewable energy production by monitoring electric equipment with connected sensors • Employment of sensors and machine learning to monitor and adapt energy consumption in buildings and industries, thereby improving energy efficiency for efficiency-driven solutions • Optimization of waste treatment and maximization of material and energy recovery from waste • AI TMS • Car energy demand prediction • Traffic jam management • AI travellers' guide • AI autopilots • Personalizd AI

(continued)

Table 10.1 (continued)

Categories of climate change measures	Methods and techniques	AI technology applications: AI frontiers
Adaptation	• Flooding • Earthquake • Landslides • Drought	• AI-driven early warning system • AI in earthquake prediction • AI rock and building crack gauger • Infrastructure stress points: Defect and risk assessment and optimization
Integration: Cross-cutting climate action measures	• Climate change as a city policy • Climate change planning and initiatives promotion • CDM • Database • Survey and research • Awareness raising and information provision • Education and training • Recognition improvement • Demand and consumption of goods and services • Supply chain management • Public safety • Edge AI (local AI) • AI campus	• Data infrastructure of AI for climate change: Big data and AI data center • Analysis of data for environmental monitoring such as mapping CO_2 and methane emissions, and biodiversity • Modern data center and cloud initiatives • AI-powered Chatbot and ChatGPT, and virtual assistants to handle citizen inquiries and provide personal information • AI human behavioral analytics and visual media art • Visual simulation • Time series analysis in demand forecasting • Regression and machine learning models for demand prediction • Optimization algorithms and their application in supply chain management • AI CCTV camera • AI systems that perform data processing and model execution directly on local devices such as smart phones, laptops or severs without cloud-based infrastructure • A platform for learning and developing AI skills, emphasizing open access and free education

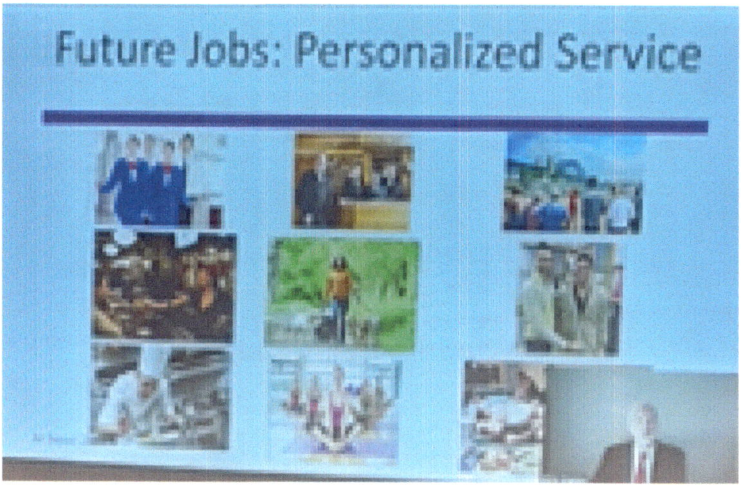

Fig. 10.1 AI future jobs: Personalized service. *Source* AI Seoul 2025: AI for a Harmonious Society, COEX, Seoul, Korea, 2025

Figure 10.1 shows AI personalized service for improving smart living and consumer behaviors to create future jobs in personalized service.

(2) AI for smart daily life to change individual consumer behavior

(a) Walmart AI wake-up call: Walmart CEO Doug McMillon has said that *AI will wipe out jobs* and reshape the workforce at the US's largest private employer. Executives say the head count will stay the same over the next few years, but the mix of jobs will change significantly.

(b) Seoul Metropolitan Government: An AI intelligent client mobility service. The AI service includes intelligent data analysis algorithms, public transportation supply data and user service system.

(c) LG: Future, together. Use of AI robots makes daily life convenient and meaningful by saving valuable time for customers.

(d) ADLINK: Personalized experience and operation efficiency through intelligent retail stores. DLAP-411 offers personalized experiences and operational efficiency through AI, Internet of Things (IoT) and automation. This digital integration helps retailers stay competitive and relevant in a fast-evolving market, meeting the demands of tech-savvy consumers.

(e) Samsung Electronics: Chatbot. This AI Chatbot uses AI-generated responses to provide immediate voice assistance for duty free retail shopping information (Figs. 10.2, 10.3, 10.4 and 10.5).

Fig. 10.2 Pursuit of AI smart life through individual mobility pattern analysis, prediction and data utilization. *Source* Seoul Metropolitan Government, Smart Life Week, COEX, Seoul, Korea, 2025

Fig. 10.3 Client service AI for creating future value. *Source* LG, 2025

10.3 Individual Element AI Technologies: Unit AI Technologies for Single AI Agents

10.3.1 FAWOO NANOTEC: AI Nanotechnology for CO_2 Reduction

Figure 10.6 shows a NanoBubble Generator for mass dissolution of carbon dioxide, as seen at the CES Innovation Awards 2025.

Fig. 10.4 Intelligent retail stores. *Source* ADLINK, 2025

10.3.2 GAPT: AI-Based GHG Reduction Solution

GAPT is an AI-based climate technology that uses a special gas analytical technique for carbon capture (Fig. 10.7).

10.3.3 Parking with: AI Parking Control System

The AI function of the Parking With system includes identification of car numbers, parking control and prevention of illegal parking (Fig. 10.8).

10.3.4 CityFarm: Low-Carbon City Farm

CityFarm is an AI system that offers an environment-friendly platform for plant cultivation, CO_2 capture and the creation of ESG value by utilizing underused urban space (Fig. 10.9).

Fig. 10.5 AI client service for retail shopping

10.3.5 EPC: Climate Tech Credits

This program provides solutions for environmental protection credits for companies that own carbon reduction technology (Fig. 10.10).

Fig. 10.6 Mass dissolution carbon dioxide for CCU. *Source* FAWOO NANOTEC, 2025

10.4 System AI Technologies

In urban areas, interconnected AI technologies facilitate the deployment of multi-AI agents in an integrated approach. This section provides some specific examples of these system AI technologies.

10.4.1 Living Lab Meeting AI Requirements of Smart Daily Life: Lifestyle Creators

(1) Hyundai Department Group: Interconnected digital IT for customers. The Hyundai Department Group suggests use of interconnected digital IT in daily life to enhance customer trust. They are commited to becoming a company with the highest level of customer trust in the fields of food, total living, distribution, and construction with digital IT.

Fig. 10.7 AI-based carbon capture for GHG reduction. *Source* GAPT, 2025

(2) Uni Tel: Connecting people, intelligent building management and intelligent visual identifiers. Uni Tel provides real-time cloud-based monitoring and alerts, enhancing productivity, and intelligent visual identifiers, correlating online and offline consumer behavior through sensors (Figs. 10.11 and 10.12).

10.4.2 AI Irrigation System

In this system, smart controllers adjust watering cycles using data from ground and weather sensors. The rectangular boxes are controller programing units; the nonrectangular piece mounted on top of the green panel are weather sensors (Fig. 10.13).

Fig. 10.8 AI driverless parking control system. *Source* Parking With, 2025

10.4.3 CRMaaS: Climate Risk Management Software

This software is used for integrated climate risk analysis, transformative risk analysis (Scope 1, 2 and 3) and physical risk management (Fig. 10.14).

Fig. 10.9 Low-carbon city farm. *Source* CityFarm, 2025

10.5 Hybrid AI Technologies

Hybrid AI technologies are a mixture of the AI unit and system technologies. Some types of hybrid technologies are:

- Edge AI inference
- AI PC
- Small scale language models
- AI workload diversity
- GPU, TPU
- Decision making framework
- AI hybrid infrastructure.

10.5.1 SK Telecom: AI Hybrid Infra for Future Extension

The AI future state looks like an integrated hybrid of edge AI inference, AI PC, a small-scale language model, AI workload diversity and a decision-making framework (Fig. 10.15).

10.5.2 Microchip: Edge AI

This program deploys machine learning models across the entire portfolio of MCUs, MPUs and FPGAs (Fig. 10.16).

Fig. 10.10 Climate tech credits. *Source* EPC, 2025

10.5.3 Dataiku: Universal AI Platform

This program offers a universal AI platform for GenAI and agents, machine learning, analytics and insights, and data preparation for AI (Fig. 10.17).

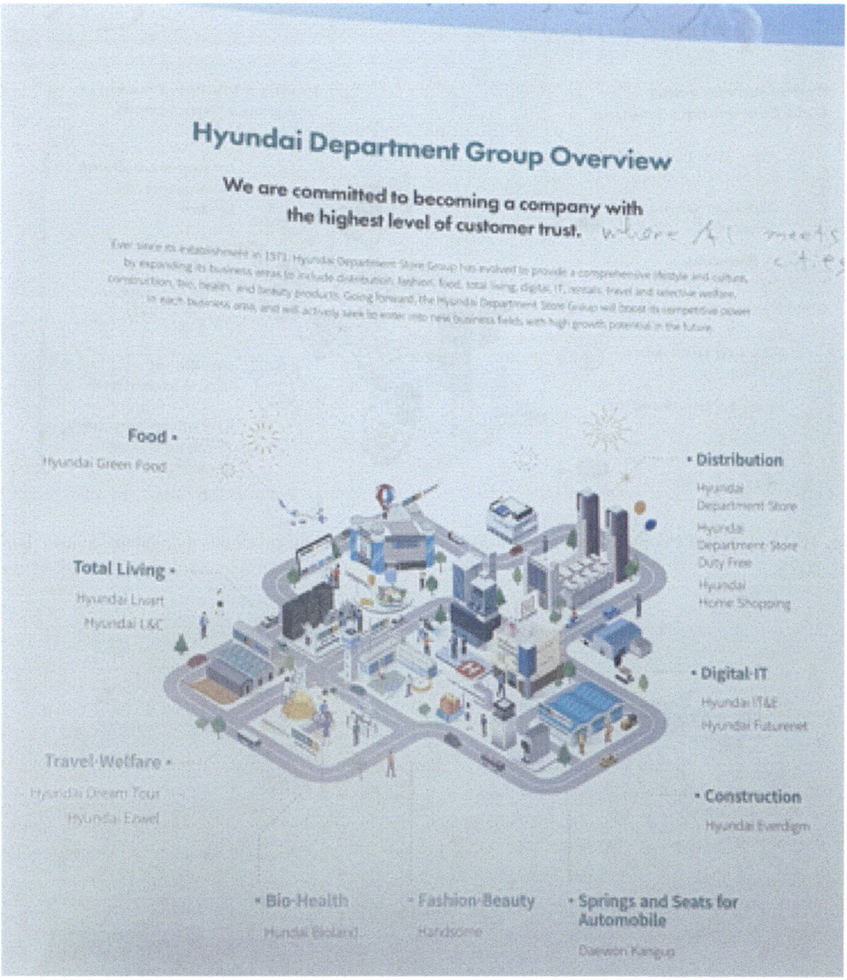

Fig. 10.11 Interconnected digital IT. *Source* Hyundai Department Group, 2025

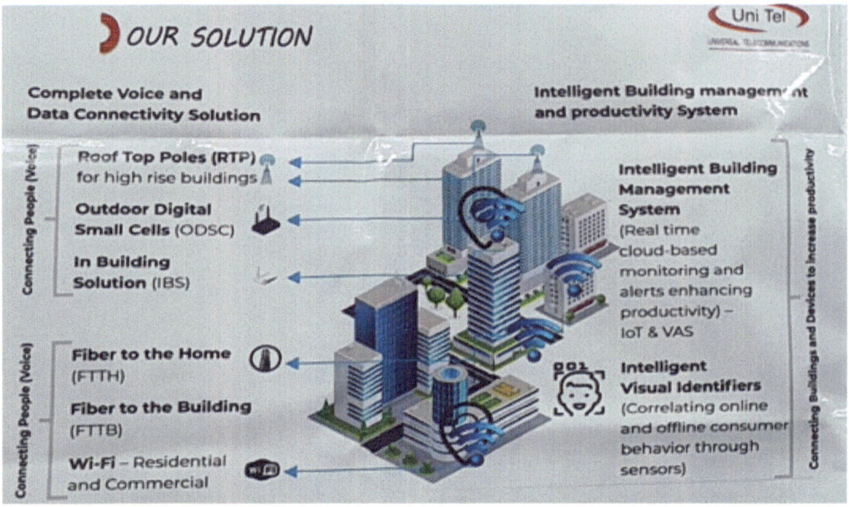

Fig. 10.12 Uni Tel:People, intelligent building management and productivity system. *Source* Uni Tel, 2025

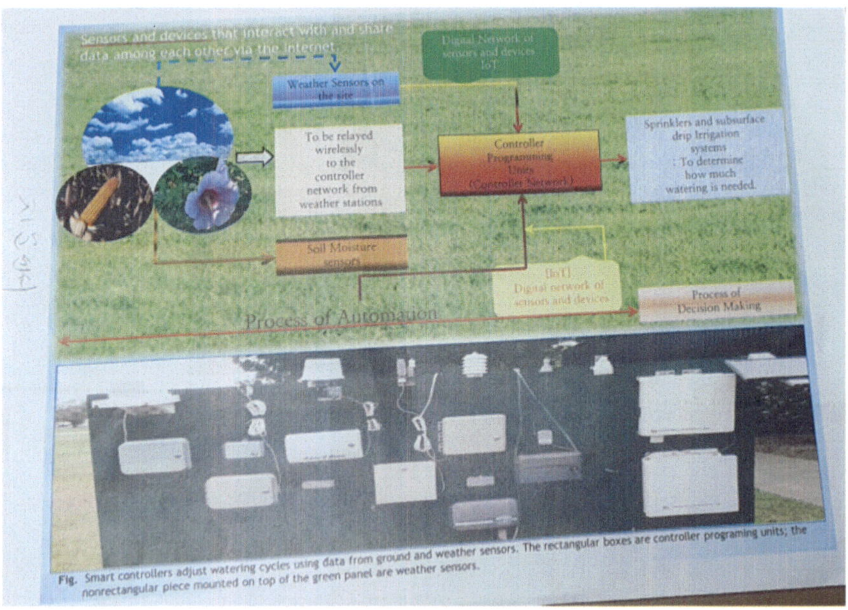

Fig. 10.13 Automated AI irrigation system. *Source* Luskin Center (undated), SMART Parks: A Toolkit, UCLA, p. 61

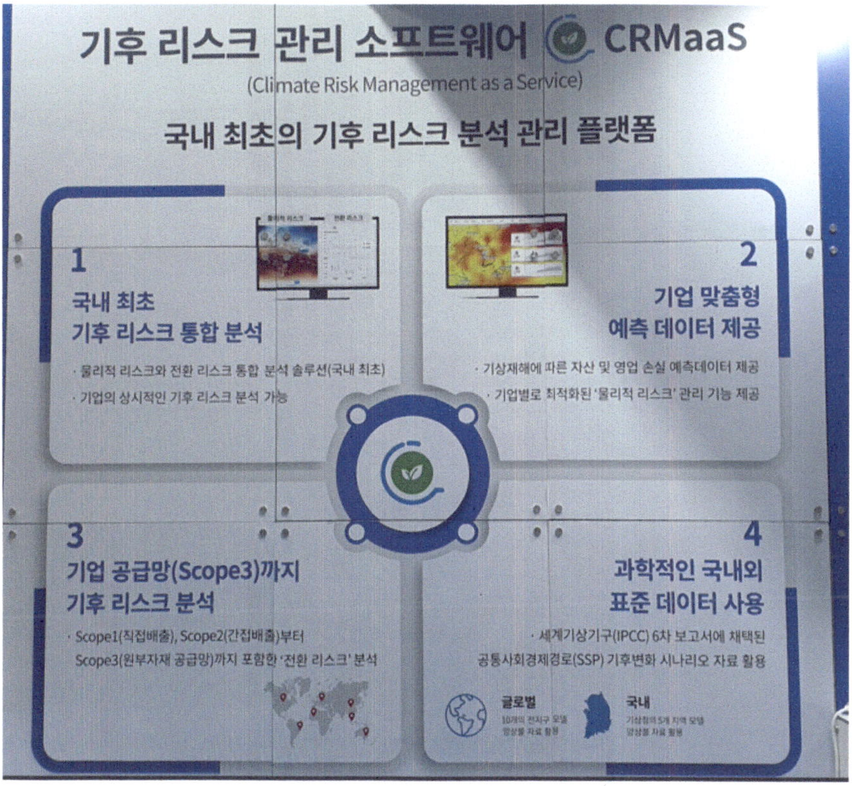

Fig. 10.14 Climate risk analysis and management platform. *Source* CRMaaS, 2025

Fig. 10.15 AI hybrid infra for future state. *Source* SK telecom 2025

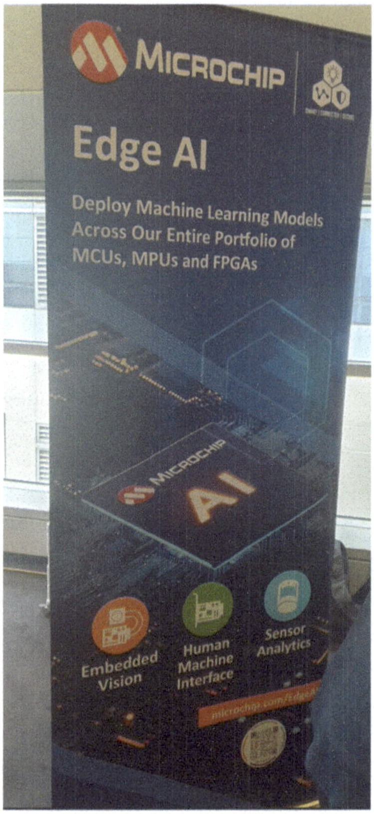

Fig. 10.16 Deployment of machine learning models for embedded vision, human machine interface and sensor analytics. *Source* Microchip 2026

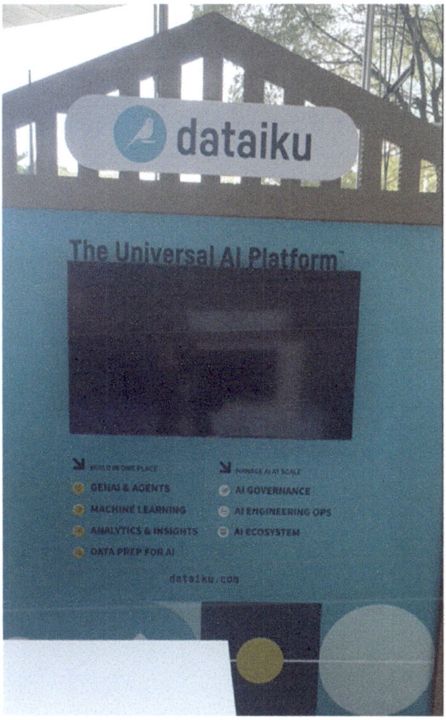

Fig. 10.17 A universal AI platform. *Source* Dataiku 2025

References

Dessler AE (2012) Introduction to modern climate change. Cambridge University Press, Cambridge, p 185

Luskin Center (undated), SMART Parks: A Toolkit, UCLA, p 61

Seoul Metropolitan Government (2025) Smart Life Week, COEX, Seoul, Korea

Chapter 11
AI Converged Startups and Business Opportunities: Fostering AI Marketing and Dynamic AI Startup Ecosystem for Climate Solutions

Kwi-Gon Kim

Abstract The AI industry is growing. Sustainability inspires success for AI jobs and growth and the startups are frontiers of emerging technologies for climate solutions. Big tech (OpenAI, KT, Google, NVIDIA, Amazon, etc.) are creating an overall AI ecosystem and AI startups provide components of the ecosystem. In Korea, although only 56% of AI startups have survived until now (Yonhap News (2025) 8 December), there are potential startups and business opportunities for engaging and exchanging with other companies, technology developers and innovators in order to create knowledge networks and to share lessons learned from success stories, as well as failures, in the field of planning, climate and environment converged with the open source AI model. This chapter takes a look at startups which are currently operational in the fields of the AI hardware and AI software industries. AI technologies are categorized, starting with the level of AI technologies for system functionalities and balancing as the underlying technical substructure of AI applications in the life cycle phase of climate-related urban project activities, described in Chap. 8.

Keywords AI industry · AI growth potential · Commercial AI models · Small-scale pilot program · Climate-tech startups · AI cloud service · Cloud-based digital twin service · Quantum computing · Data visualization platform · Blockchain · Tech-giants · Green hydrogen · AI-enhanced urban living · AI business model · Google campus

11.1 Growth Potential: Data Centers and AI

The history of emerging technology development, which provided the foundation for the evolution of AI startups during the period from 2006 to 2025, is illustrated in Fig. 11.1.

K.-G. Kim (✉)
Seoul National University, Seoul, South Korea
e-mail: kwigon@snu.ac.kr

K.-G. Kim (ed.), *AI Transformation of Cities for Climate Solutions*,
https://doi.org/10.1007/978-3-032-21940-4_11

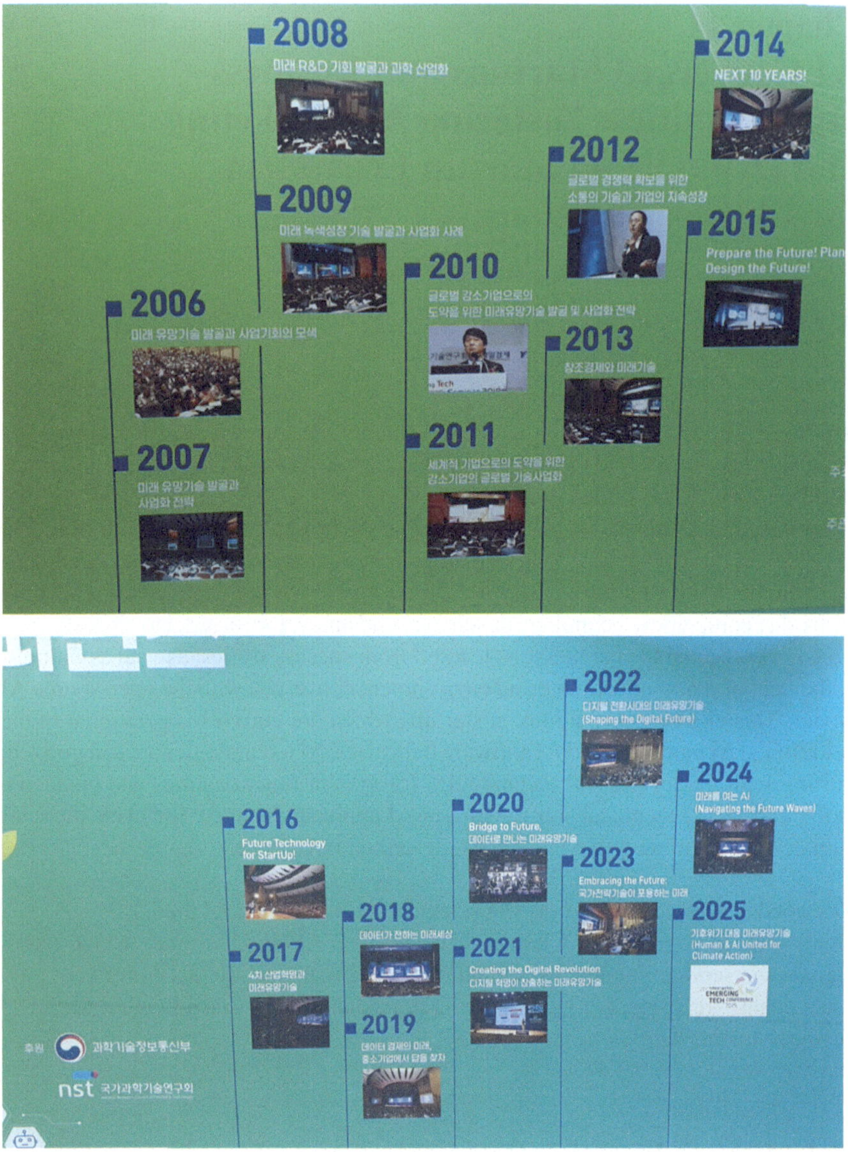

Fig. 11.1 History of emerging technology development. *Source* Ministry of Science, Technology, and Information and Communication (2025)

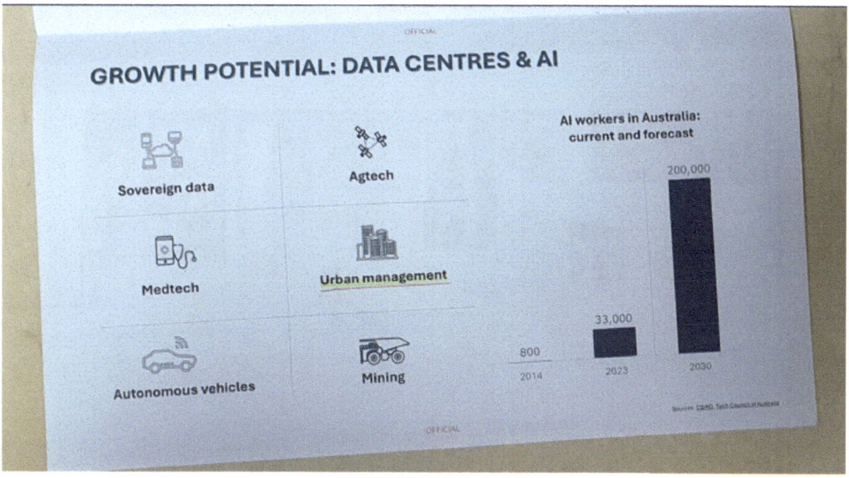

Fig. 11.2 Growth potential of AI urban management. *Source* Australian Trade and Investment Commission (2025)

Until recently, AI tech giants have focused on future new AI industry such as research and development of Open AI ChatGPT, Google Gemini, Perplexity, Chinese DeepSeek, Kimmy K2, UAM and robots and autonomous vehicles. As can be seen in Fig. 11.2, AI urban management is rapidly growing and provides a range of opportunities for AI R & D. It is now time for startups to invent AI models which are free to access outside of commercial offerings.

Workers fear being replaced by AI—or just by someone who knows how to use it. At companies big and small, the technology is starting to cost workers their jobs if their bosses believe they aren't embracing it fast enough. Meanwhile on Wall Street, this week the AI boom finally seemed to get a reality check (The Wall Street Journal, 9 November 2025).

Figure 11.3 shows that startups need to advance their skills in order to influence the future of AI. Even though big tech companies have developed an overall AI ecosystem, startups should provide support as components of the ecosystem, and advance their skills in order to help shape the future of AI.

11.2 Climate Value Up of AI Startups to Be Developed: What's Ahead for the AI-Driven City

Climate-change focused AI cities should establish small-scale pilot programs to reduce the hazards associated with large-scale technology installations. Cities can experiment with new AI technologies in a manageable context by launching restricted

Fig. 11.3 Startups dedicated to shaping the future

experiments and allowing for thorough monitoring and required tweaks to improve their performance before a larger rollout.

This incremental technique allows urban planners and policymakers to detect possible problems, collect useful data and fine-tune technologies based on real-world performance and community feedback. Successful pilot programs in places such as Paris and Lyon, which underwent thorough testing and strategic planning, show that this strategy is viable and beneficial. These pilot efforts not only aid in fine-tuning technologies, but also in establishing public trust and stakeholder buy-in.

By demonstrating actual outcomes and benefits on a smaller scale, communities may successfully address concerns and demonstrate the positive impact of AI-driven solutions. This strategy also encourages innovation because insights acquired from pilot initiatives can help create more robust and scalable solutions.

'Climate startups' (or climate-tech/climate-tech startups) are really diverse. Summarized lists of innovative potential computer programs in urban planning and development, which can be utilized by AI startups or tech giants are presented in the next sections.

11.2.1 AI Tech Giant Programs and Projects

- AI models and algorithms
- ASI (Artificial Super Intelligence)
- Robots
- AI agents
- Unmanned mobility (drones, driverless cars, etc.)
- Chatbot
- GPU, TPU
- Cloud computing (AI cloud service)

- Algorithms (land use planning, etc.)
- Machine learning (deep learning, etc.)
- AI data center
- Data and AI platform management
- Communication: 5G/6G
- Cloud-based digital twin service
- Quantum computing
- ESS.

11.2.2 AI Startup Programs and Projects

- Metaverse
- AI-based digital twin solution platform
- 3D printing
- AI PC
- Intelligent AI system of water supply and sewage
- AI system for photovoltaic lighting system abnormal detection and power generation prediction
- Online delivery platform
- climate disaster and risk management
- Smart grids
- Blue-green infrastructure
- Spatial intelligence: Space computing (space information platform)
- Web standard technology
- GIS software
- AI soft power
- Logistics management
- IoT-based facility management system
- Construction and maintenance
- Data visualization platform
- Sustainable AI supply chain
- Demand forecasting and procurement
- Evaluation of AI system functionalities and balancing education
- Risk analysis and value chain strategies
- Data analytics
- AI-based consumption survey tools
- Blockchain (tokenization)
- Climate tech nexus
- AI digital smart home security solution
- Integrated parking lot system control
- Bus information terminal (BIT)
- AI video generation technology
- Carbon capture and removal

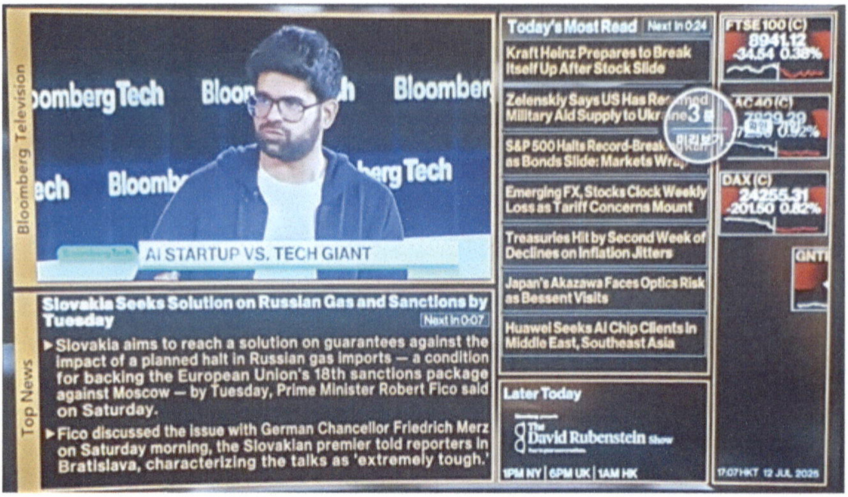

Fig. 11.4 The AI race between AI startups and tech giants. *Source* Bloomberg Tech (2025)

- Clean energy/hydrogen
- Climate Fintech
- Carbon-negative cement
- Digital door lock.

An AI race may materialize between AI startups and tech giants to develop the programs and projects listed (Fig. 11.4). In fact, many tech giants including Google, Amazon, Meta, and Tesla began as startups but have become global corporations.

11.2.3 Key Trends Among Climate Startups

The key trends among climate startups can be summarized as follows:

- Carbon removal is scaling
- Green hydrogen is gaining traction
- AI-managed energy storage and smart grids are being innovated
- Data-driven climate actions are becoming more important for companies to measure and manage their carbon footprint.

11.3 Future Opportunities for Startups Generated by AI

In terms of climate city planning, there are additional areas for startups, which could be explored in the future:

- Startups of AI and real estate development: e.g., Apartments using green hydrogen produced from wind power generation farms
- Startups of AI and climate impact assessment
- Startups of AI and climate risk management
- Startups of AI and ecological impact assessment
- Startups of AI and climate change mitigation
- Startups of AI and climate change adaptation
- Startups of AI sustainability cost–benefit analysis
- Startups of AI-enhanced urban living (Fig. 11.5).

Figure 11.5 shows various AI ideas of urban living including sleeping, growing plants, studying and playing musical instruments.

In general, future prospects for AI startups appear very strong. The field should put more emphasis on startups for climate and energy optimization, combining AI, climate domain expertise and real workflows. However, there are some challenges:

- Capital requirements are rising due to the cost of climate servers and data and model training

Fig. 11.5 AI ideas of urban living. *Source* KT, K-AI, 2025

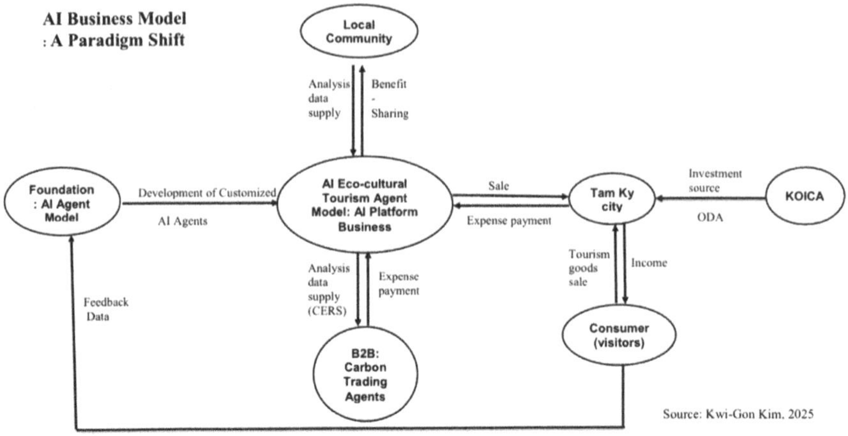

Fig. 11.6 A proposed KOICA startup business model

- Climate startups will need proprietary data, domain specialization, marketability, efficiency or unique AI agent model architectures.

Efforts should be made to create new markets for AI climate startups. Figure 11.6 proposes an AI business model as a paradigm shift. At the center of this model, AI startups could play the role of an innovation hub that could contribute to devising an AI agent model platform for business.

Figures 11.7 and 11.8 share lessons learned from success stories about Google campus for startups. We can discover how leading women startup founders from Google Founders Academy APAC program are using their innovative AI technology to shape an inclusive future.

Fig. 11.7 Google campus for startups, Seoul, Korea

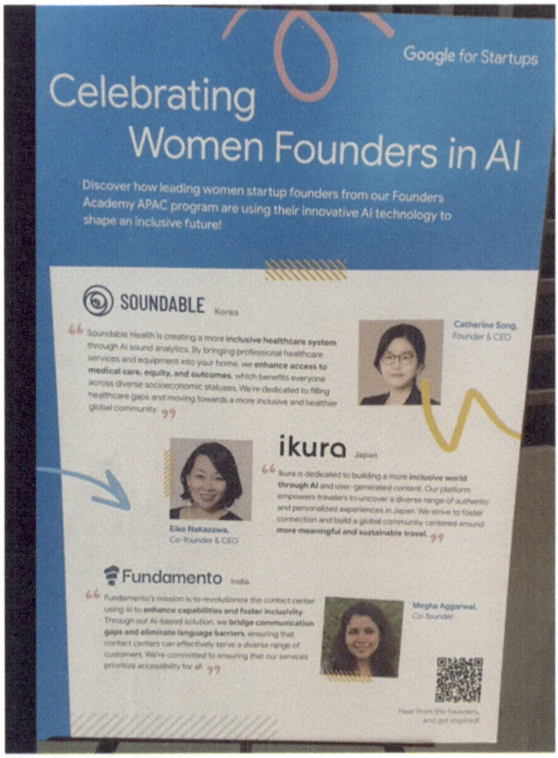

Fig. 11.8 Success stories celebrating women founders in AI. *Source* Google Campus for Startups (2025)

References

Australian Trade and Investment Commission (2025) Australia's data center landscape: current and future potentials for AI and energy

Bloomberg Tech (2025) AI race between AI startup and tech giant, 12 July

Google Campus for Startups (2025) Success stories celebrating women founders in AI

History of Emerging Technology Development (unpublished)

Ministry of Scince, Technology, Information and Communication (2025)

The Wall Street Journal (2025) 'Workers fear being replaced by AI', 9 November

Yonhab News (2025) 8 December

Part V
Applications: Incorporation of AI Technologies into Cities

Chapter 12
AI Virtual City Proposals for Sustainability: AI Imaging Technologies Including Virtual Twins

Kwi-Gon Kim

Abstract Cloud and data warehouses are connected through data visualization. Even though the urban planning and development process is technically very complex, two significant components in designing the process are prediction and optimization for decision making both of which can be improved by an AI platform with tangible impacts. The use of an AI platform can aid planners and developers in modeling future change based on the indicator approach, and optimizing smart and sustainable eco-city solutions within the bounds of data availability. Inferential AI models can be used to predict future city images and solutions for sustainability based on existing data. Data and AI scientists, optimization experts, optimum system experts and urban planners could work together in designing such inferential models for AI virtual cities. These types of cities could provide a carefully optimized set of virtual urban forms and spaces intended to meet smart and sustainable growth indicators. The virtual spaces could be designed and allocated on the city maps to directly reflect the sort of AI urbanism which involves an idealization of AI technologies. This model can be tested in real-world urbanism. This chapter discusses some issues and solutions in virtual projects establishing sustainability, resiliency, diversity, connectivity and circularity of cities by AI means from a smart and sustainable urban growth perspective. It also reviews some AI visual simulation techniques and algorithms.

Keywords Virtual twin · Inferential AI models · AI virtual cities · AI visual simulation technique · Visual representation · Synthetic visual environment · Visual world · Real world · Metaverse · Virtual reality · Spatial intelligent simulation · Virtual city network · Virtual image · Urban behavior monitoring

---

K.-G. Kim (✉)
Seoul National University, Seoul, South Korea
e-mail: kwigon@snu.ac.kr

K.-G. Kim (ed.), *AI Transformation of Cities for Climate Solutions*,
https://doi.org/10.1007/978-3-032-21940-4_12

12.1 Advantages of AI Visual Simulation Techniques in City Climate Planning

AI technology can enable a great leap forward in the ability of city climate planning to offer visual representation of the ways in which the climate has changed, and will continue to change, based upon planning decisions. Until recently, however, the majority of climate planning ideas developed with AI visual simulation techniques have been conservative, and oriented to past historical adaptation practices.

In city climate planning, most changes have taken place with reference to mathematical climate impact modeling. But the introduction of AI technology can enable a great connection between climate change, change of land use and settlement patterns and visual simulation techniques.

AI produces simulated images or scenes to reduce the need for real-world data, and uses synthetic visual environments to train models and then transfer them to real-world tasks.

12.2 Use of Urban Data Analytics and Digital Twins

Spatial models, from analytical simulations of local neighborhoods to large-scale systems of cities, are crucial in urban data analytics.

A digital twin platform helps real time-based virtual models, AI-based digital mapping, 3D modeling and simulation (Fig. 12.1). Figure 12.2 shows real-time AI analysis, a centralized monitoring and control platform, an intuitive dashboard to visualize potential risks, and leveraging AI technology to increase workplace safety.

The digital twin in Fig. 12.3 is used for data collection and AI algorithm solution in the virtual world instead of the real world to provide more value and functionality through the connection of knowledge and information with industrial sites. Real-time 3D site monitoring helps to assess what actually happens on the site.

12.3 Use of the Metaverse

The term 'metaverse' generally refers to a collective, persistent digital universe where people can interact, work, play, create and socialize through avatars—often using technologies like virtual reality (VR), augmented reality (AR), blockchain and advanced 3D simulations (ChatGPT 2025).

Seoul Metropolitan Government has created a metaverse cyberspace. Figure 12.4 presents a future image of solar energy grid systems through virtual reality simulation.

Fig. 12.1 A digital twin platform for environment, energy and smart city. *Source* DT, 2025

12.4 Use of Spatial Intelligent Technologies

We live in an era of simulation with spatial intelligence. Figure 12.5 provides examples of simulations in spatial intelligence which are extremely valuable in technical fields related to AI.

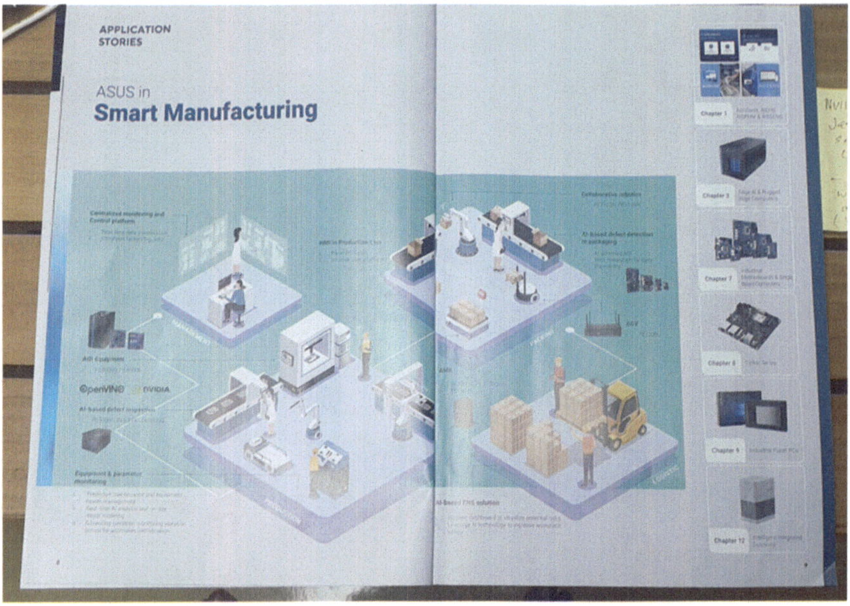

Fig. 12.2 Application stories of AI-based imagination. *Source* ASUS, 2025

12.5 Image of Climate-Smart and Sustainable City Network

A virtual image of a CO_2-related smart city network and relevant AI equipment (ASUS 2025) is shown in Fig. 12.6. It is based on a connected system of smart energy, water management, smart lighting, smart transportation, smart parking and waste management.

12.6 Project Proposals for AI-Powered Virtual Worlds

AvaCapo (2025) has developed several human behavior interaction system models by using spatial intelligent technologies as follows:

- Project 1: AI-powered urban behavior monitoring (pilot program)
- Project 2: AI avatar tourist information screens
- Project 3: Virtual fitting room for traditional Vietnamese clothing
- Project 4: AI cultural photobooth—Vietnamese heritage edition
- Project 5: City AR experience—mascots and cultural IP in augmented reality.

These proposals present a set of technology-based projects aimed at supporting city infrastructure, public services and tourism. The solutions include both functional

Fig. 12.3 Digital twin solution: 3D construction site monitoring. *Source* VIRNECT, Twin, 2025

systems, such as urban behavior monitoring, and interactive public installations based on AI and XR technologies, including digital avatars, virtual fitting rooms, cultural photobooths and AR experiences involving local mascots and cultural content.

Fig. 12.4 Virtual Reality (VR) related to metaverse cyberspace showing solar energy grid systems, Seoul, Korea. *Source* Seoul Metropolitan City, Smart City Exhibition, Citizens' all, 2024

Fig. 12.5 Illustrations of spatial intelligent simulation. *Source* SPRI, Software Conference, COEX, Seoul, Korea, 2025

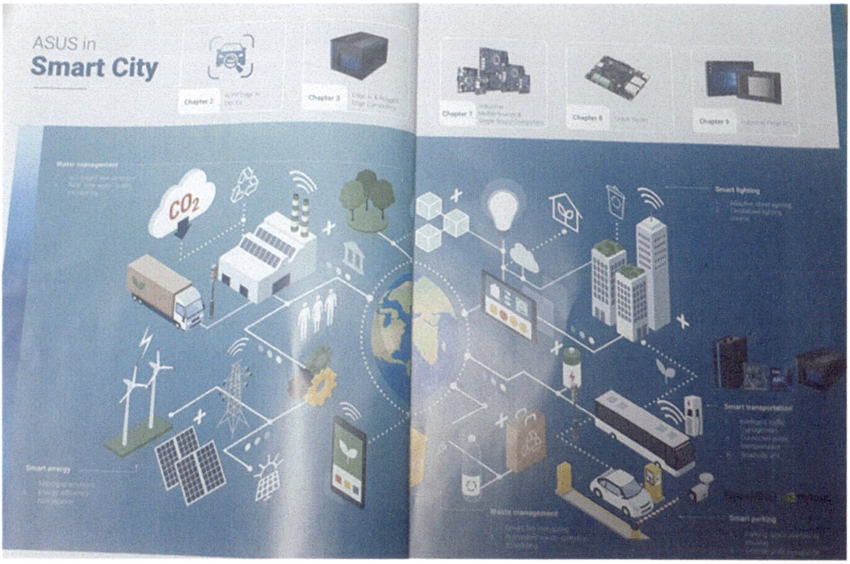

Fig. 12.6 Virtual image of a connected city system. *Source* ASUS, 2025

References

AvaCapo (2025) Human behaviour interaction system models, unpublished
ChatGPT (2025) website

Chapter 13
Applying AI in the Urban Context to Scale Green Finance and Carbon Markets for Sustainable Urban Development

Michael Schmitz and Catherine Atkin

13.1 Introduction

13.1.1 The Need for New Solutions

Cities and local governments face severe challenges both in meeting the needs of their residents for core services and also taking action to ensure the long-term health of people and the environment. While the debate over the societal impacts of artificial intelligence (AI) grows increasingly heated, the use of AI to support sustainable, climate-positive urban development is being tested by localities across the world. Whether it is through the development of AI-driven solutions for cost-effective monitoring, reporting and verification (MRV) that supports the development of baseline measurements needed for the issuance of green bonds and the generation of trusted carbon credits, or the use of AI for predictive analytics to assess climate risks, the deployment of these algorithms and tools offer great promise for those seeking to unlock green finance and catalyze carbon markets.

M. Schmitz (✉) · C. Atkin
Stanford Codex Law x Climate, Stanford, California, USA
e-mail: michael.schmitz@codex.stanford.edu

C. Atkin
e-mail: catherine.atkin@codex.stanford.edu

K.-G. Kim (ed.), *AI Transformation of Cities for Climate Solutions*,
https://doi.org/10.1007/978-3-032-21940-4_13

13.1.2 Cities Begin Leveraging AI, but a Gap Remains for Urban Development

Localities across the world are experimenting with the use of AI to address hard-to-solve, complex challenges, such as issues in urban transportation and energy systems, with some early successes. There are a growing number of examples of applications of AI being used to assess both asset classes and specific processes in urban energy, transportation and health systems. The types of applications being deployed are wide ranging, from computer-vision to sensor-data analysis programs and large language models (LLMs).

In Singapore, AI is being used to analyze satellite and street-level imagery to allow for real-time monitoring of building energy performance, improving auditability and compliance with local climate and energy requirements (Biljecki 2025). Private firms in New York City, USA, are using AI to help regulated building owners comply with new building and efficiency disclosure laws (Burns 2025). In Oslo, Norway, AI is being deployed to analyze sensor data for forecasting energy demand from the city's charging system, as it faces rapid growth of demand from new EVs, and needs to stabilize the grid and improve reliability (WEF 2018). Barcelona, Spain, is utilizing AI in the transportation context to predict and manage congestion from cars and reduce transportation emissions (Mayors of Europe 2025). In Bengaluru, India, AI models are being used to assess upcoming extreme heat events and help prepare for these events and their public health impacts (Girgekar 2025, ATACH 2024). In San José, California, city leaders are showing a willingness to take a strategic approach to AI deployment, including using AI for bus route optimization (Edinger 2025).

San José's Chief Innovation Officer, Stephen Caines, underscored that city's ambition:

> San José is using the enormous potential of AI to grow the economy, empower the community and spur innovation. We have a once in a lifetime opportunity to use the power of AI for the common good and San José, as the capital of Silicon Valley, is all in on realizing its potential (S. Caines, personal communication, December 2025).

It is important to note that these early interventions are not focused on sustainable urban development and can be characterized as individual responses to specific challenges and opportunities rather than as the result of integrated long-term planning efforts and investment in the AI infrastructure development needed to support the scaling of green finance. As important as such demonstration projects are in the design and development of smart, AI-enabled systems, there remains a significant gap between initial piloting of AI in cities, and the system-wide deployment of solutions that will be required to fundamentally address the energy and climate challenges in the urban context by strengthening the digital data infrastructure required to unlock financial flows. Bridging this gap will require strategies and investments that build the AI infrastructure in a way that connects fragmented, heterogenous city databases, and integrates them to enable both real-time analysis by operators and policymakers as well as automation of processes to increase systems efficiencies and effectiveness required to unlock public and private capital flows.

13.2 The Potential for AI to Catalyze Urban Green Finance

The need to take concerted action to address the climate finance gap features prominently in global dialogues. According to the Climate Policy Initiative in the Global Landscape of Climate Finance 2025, 'to reach the 1.5 °C scenario, a fivefold increase in annual climate finance is required, reaching an average of $7.4 trillion needed each year through 2030' (Naran et al. 2025).

The building of integrated energy and climate-related systems, supported by AI algorithms and applications, is a necessary prerequisite to scaling green finance for many of the carbon reduction, green energy and climate-related, early-stage projects that face financial, technical and perceptual risks. In this regard, the real promise of a well-functioning, intelligent city infrastructure is that green financing will be able to flow in the right directions and at the scale needed to take early-stage projects and create market leading companies able to transform urban energy, transportation and health systems.

It is critical to have a focus on energy and climate strategies in the world's urbanized regions, which account for two-thirds of global energy consumption (IEA 2021) and three-quarters of greenhouse gas emissions (IPCC 2022). Projections for continued new urban development in the next decades will be concentrated in the global South (World Bank 2025), adding to global population growth which is estimated to reach 10.2 billion people by 2100 (UNDESA 2024). This type of growth will require systems that can simultaneously produce significantly more energy, in more regions, and significantly reduce emissions from new and existing energy production and power generation sources. This new low-carbon development will need financing in an era marked by decreasing public trust in the public institutions responsible for guiding such development (OECD 2024).

Given the current landscape, AI shows genuine potential for addressing some of the most difficult challenges and blockers to the deployment of green bonds and the development of carbon markets. From the perspective of developing trusted green-finance systems, there are several clear uses for AI in cities to enhance the efficacy, transparency and scalability of public and private green development in urban areas: (1) the establishment of measurement, reporting and verification (MRV) systems of emissions and energy usage; (2) the creation and maintenance of baseline models of emissions and energy usage; and (3) the development of dynamic predictive models for auditing and compliance. A deeper look at the required elements for issuing green bonds and carbon markets and the application of AI shows these elements to be of primary importance, and, therefore, the subject of this analysis.

13.2.1 Green Bonds and AI Applications

While green bond issuance is growing, with estimates of a $1 trillion market size in 2022, (S&P Global Ratings 2025) forecasts suggest that this will only amount to

12–15 percent of what is needed to address the climate crisis. These same reports identify data quality as a barrier to growing the market (S&P Global Ratings 2024).

Localities pursuing strategies to finance carbon reduction projects are increasingly looking to issue 'green bonds' for environmentally beneficial infrastructure projects. A leading proponent of green bond financing, the International Capital Markets Association (ICMA) defines a green bond as 'any type of bond instrument where the proceeds or an equivalent amount will be exclusively applied to finance or re-finance, in part or in full, new and/ or existing eligible Green Projects' (ICMA 2025).

The Green Bond Principles (GBP), outlined by the ICMA, is a global voluntary collaborative to scale this form of green financing. The GBP issues guidelines for issuers of these bonds to promote transparency, disclosure and integrity in the green bond market. There is robust dialogue in the research and finance community about the opportunity to leverage AI tools and capabilities to help issuers more quickly and effectively adhere to each of the components of the Green Bond Principles and ensure successful project outcomes (Jia et al. 2025; Le Fur 2025).

Table 13.1 provides an overview of the ways in which AI can add significant value in the green bond sector. Notable are the opportunities for measurement, reporting and verification (MRV), baseline modeling, and dynamic auditing and compliance.

13.2.2 Carbon Markets

Developers of carbon mitigation projects have been largely dependent on the small, fragmented voluntary carbon markets (VCMs) as sources of investment. These markets often rely on manual, non-standardized processes for issuing and trading low-quality carbon credits and have been the subject of intense criticism and allegations of 'greenwashing' resulting in negative impacts on the speed and scale of issuance and trading, as well as the overall confidence in VCMs (Gołasa et al. 2025; Probst et al. 2024).

The emergence of compliance carbon markets in the form of emissions trading systems across the globe and the imposition of the Carbon Border Adjustment Mechanism (CBAM) in the EU create an imperative to ensure robust data. The use of AI combined with digitization of the credit generation process offers great potential for significant market growth, increased liquidity and greater confidence and trust in carbon markets.

This new reality has led to a robust discussion of ways to create more confidence in the data value chain focused on measurement, reporting and verification. AI and blockchain technologies are powerful tools and are being integrated into solutions (Kumar et al. 2023).

Table 13.1 Green bond elements and AI applications

Green bond issuance components ([a])	Focus for issuer	Opportunity to leverage AI
Use of proceeds	Issuer to clearly define the environmental benefits of the project, capable of being assessed and quantified where feasible	Baselining: AI can quantify a project's likely environmental benefits by drawing upon dynamic climate modeling and automated digital asset tagging, as well as ensuring alignment with eligibility requirements
Process for project evaluation and selection	Issuer to describe sustainability objectives, process for determining fit, complementary information	Baselining: AI can use geospatial data and predictive analytics to assess future project fit to defined sustainability objectives
Management of proceeds	Issuer to credit internal sub-account, and appropriately track, encouraging 3rd party verification	MRV & Compliance: AI can process issuer fund flows in an ongoing, real-time manner, use machine learning to detect anomalies, and produce a digital audit trail
Reporting	Issuers to make and keep up-to-date information on the use of proceeds, renewing annually until full allocation	MRV & Compliance: AI can automate the synthesis of sensor and satellite data, accurately estimate performance and ensure transparent reporting

[a] International Capital Market Association (ICMA) (June 2025) The Green Bond Principles

13.3 Key AI Applications

The discussion in Sect. 13.2 demonstrates the power of AI to address key requirements for the needed expansion of the green bond and carbon markets outlined in Tables 13.1 and 13.2. Three applications of AI are particularly relevant to the growth of both green bonds and carbon markets: Measurement, reporting and verification (MRV), baseline determination and modeling, and dynamic predictive models for auditing and compliance. We now turn to a more detailed examination of each of these applications.

13.3.1 AI-Powered Measurement, Reporting and Verification

One of the most direct and powerful uses of AI is for MRV in the context of dynamic, hyper-complex urban settings, like city transportation systems. A growing number of AI-enabled systems are already in place to ingest multiple datasets from sensors,

Table 13.2 Overview of carbon credit generation

Core carbon credit generation stages [a]	Data requirement	Opportunity to leverage AI
Project scoping & baselining	Establish accurate baseline and parameters for additionality	Baselining: AI analysis of geospatial data for pattern detection and estimation of baseline
Methodology & design processes	Align credit design with registry protocols to ensure qualification	Baselining: Use of digital twins and natural language processes to automate project design
Monitoring	Track the issuance and retirement of credits	MRV & Compliance: AI analyzing satellite, IoT and computer vision data for real-time tracking
Auditing	Ensure auditability of full credit lifecycle	MRV & Compliance: Use of ML algorithms to automate detection of anomalies
Issuing & registering	Proper issuance of credits and recording with aligned registry	MRV: Use smart contracts to automate the serializing and recording of credits

[a] Washington, DC: World Bank (2022) Digital Monitoring, Reporting and Verification Systems and Their Application in Future Carbon Markets. [License: CC BY 3.0 IGO]

satellites and self-reporting from entities in compliance-reporting regimes, and integrate these data sources into actionable intelligence by system operators. Innovative designers are deploying a number of different technology strategies to make breakthroughs in this context.

AI-enabled MRV systems are being used for monitoring vehicle usage in urban centers for a variety of uses including congestion pricing, driver safety and intelligent transportation systems (Fig. 13.1).

Convolutional Neural Networks (CNNs) are being used for real-time processing of data and for detection of anomalies. These networks can be used for image processing and Long Short-Term Memory (LSTMs) for time-series data analyses, allowing for the integration of data from IoT sensors, ground monitors and satellites. With these tools in place, it is possible to have automatic detection of discrepancies between actual measured emissions and reported emissions, for example, and much more quickly identify unreported emissions releases and process and/or equipment failures (Al-Fuqaha et al. 2024).

MRV and Green Finance

The integrity of green financial systems, including green bonds, climate tech investments, carbon markets and carbon reduction credits, is largely dependent on ensuring

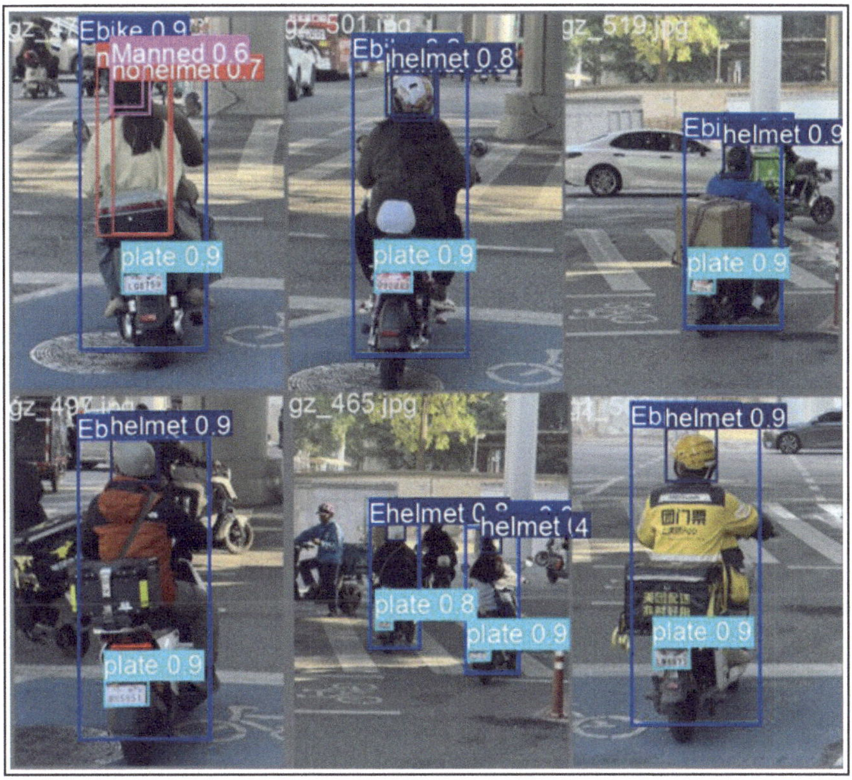

Fig. 13.1 Using computer-vision (object detection) models in electric bike safety detection. *Source* Zong et al. 2025

trust in the underlying units of economic value and the process of their creation. Building and maintaining an integrous green financial system requires rigorous verification, transparent monitoring and confidence in the reporting of associated emissions reductions. With respect to each of these elements, AI could help advance green finance by moving from the current periodic, manual, self-reporting reality to a continuous, automated, publicly transparent system that instills confidence and enhances liquidity in these markets. For example, tree canopy and biomass cover can be very effectively monitored with AI analyses of satellite imagery combined with ground truth sensing. Such processes can be easily automated and provide verification at a fraction of the cost of current approaches (Pachama/Digiqt 2025).

When it comes to issuing a green bond or creating tradable carbon credits, effective MRV systems are key to maximizing value, but they have yet to be fully leveraged in the urban context. Geospatial (GIS) mapping has long been deployed in the context of urban information systems. AI is now enabling GIS systems to greatly improve the utility of aerial imagery (e.g., satellite and drone) to delineate jurisdictional boundaries automatically, and, relatedly, dynamically monitor land-use changes. Such ML

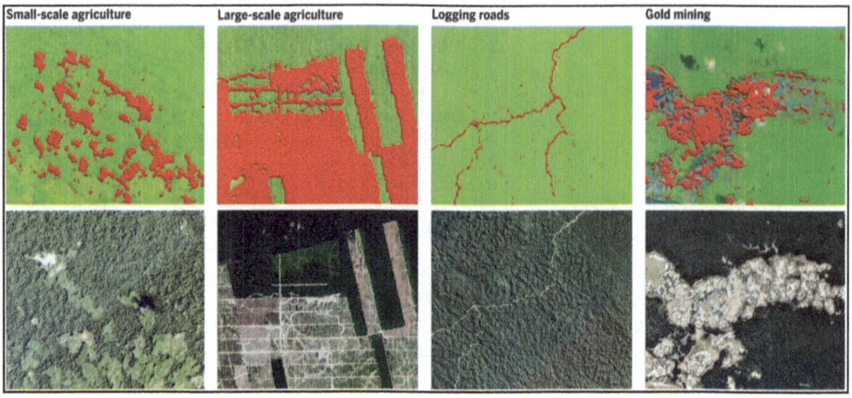

Fig. 13.2 Differentiating different types of deforestation from satellite imagery. *Source* De Wilde 2023

models allow for the analysis of aerial imagery to much more accurately quantify the stock of carbon embedded or emitted in forests or soils, and allow for much more rigorous quantification of nature-based projects (Fig. 13.2).

Machine learning systems can also provide the basis for highly accurate predictive modeling and forecasting applications. Such deployments can effectively forecast both emission trends and potential spikes based on historical data. With these systems in place, operators can more effectively implement proactive measures to take corrective actions and prevent breaches of regulatory limits, while using predictive maintenance to ensure systems function better and longer.

One particularly useful potential application of an AI model for green finance would be establishing causal relationships between specific strategies, like an industrial or natural process, and predicting the resulting emissions reductions. Such a model would enable the operator and/or monitor to more effectively detect leakage and track displacement of an emitting activity from a monitored to unmonitored area.

Another significant potential application would use AI to validate MRV data and integrate that data with smart contracts built into blockchain base layer protocols to allow for full automation of the issuance and transfer of carbon credits. In this use case, the verification of an emissions reduction event by an AI algorithm would trigger a smart contract to execute the transaction of a relevant carbon credit. Such a system could deliver secure, immutable and fully automated verification to the credit process (Islam 2025).

13.3.2 AI-Powered Baseline Determination, Modeling and Projections

In addition to their utility as enabling systems for continuous MRV, AI algorithms offer great potential for addressing the difficult challenges of creating and maintaining credible emissions reduction scenarios and plans.

AI algorithms can help establish much more accurate baseline models as a result of the ability of these programs to ingest and analyze extremely large and disparate datasets of geospatial, land-use, climate and other relevant data and produce much more granular, accurate projections of likely baseline emissions than existing methods that rely more heavily on manual collection and analysis of data (Soubry et al. 2024).

Application for Green Finance

In most carbon credit trading systems, additionality is a key element of determining the value of credits, measuring the difference between what would have occurred 'but for' the carbon reduction or sequestration intervention and what is actually measured as a result of the project. Establishing an accurate baseline, or counterfactual scenario, is, therefore, critical to ensuring the value of each created carbon credit regardless of program or project design, and allows for more accurate projections for the project or process being considered. Similarly, the accurate measurement of the resulting emissions of a carbon-reduction project is key to assuring the created credits have delivered the reductions that the project developers and operators purport. In both aspects, AI can improve these processes and help build trust and confidence as a result (Guizar-Coutiño et al. 2020; Smith et al. 2025).

Using more granular and accurate baseline projections, it is possible to increase the accuracy of projections of future production of carbon reductions from the project or process in question. Using AI algorithms integrated into such forecasting would also enable the inclusion of a wide range of variables, measuring, for example, weather patterns, risk of disease to nature-based assets, soil health and other elements affecting the impact of the scheme (Smith et al. 2025).

With more accurate baseline modeling and forecasting of credit value, assessing project-related risks would improve, helping inform project investors and credit purchasers with insights on regulatory uncertainty and the risk of premature release or leakage of captured or stored carbon. For example, cities can use green bond funding for a municipal fleet electrification project, where AI would analyze years of fuel purchasing, mileage and vehicle-efficiency data for an existing internal combustion engine (ICE) municipal fleet to calculate an exact annual emissions baseline that an electrified city fleet would be required to exceed (Deloitte Canada 2023).

The use of digital twins is another very powerful application of an AI model in this context. Digital twins, or virtual replicas, of assets can be created for a wide range of applications including industrial and natural systems. A digital twin can

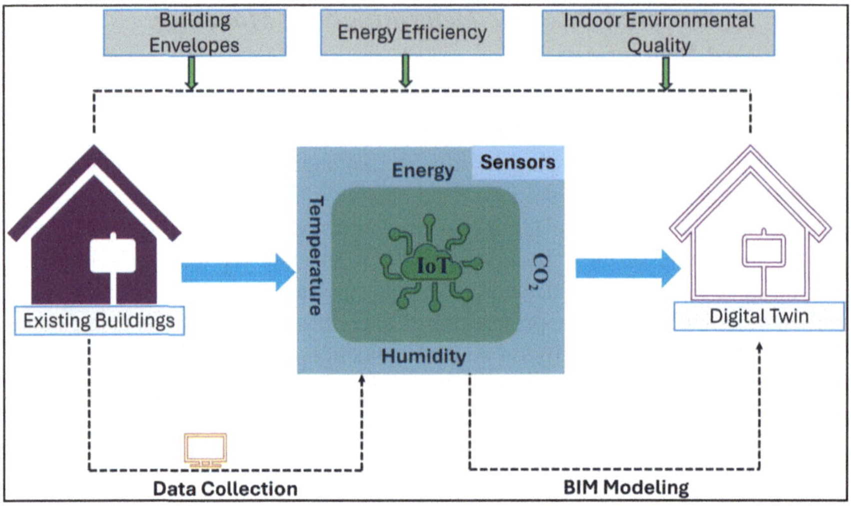

Fig. 13.3 Schemata for using digital twins for existing buildings. *Source* Venkateswarlu et al. 2025

ingest real-time data from the real-world entity and allow for simulations to be run near-real time, enabling proactive assessments of the carbon and energy impacts of operational changes. In this deployment, MRV is transformed from a reactive tool for validation to a proactive tool for risk profiling and assurance. Green bonds, focused on urban climate-resilient housing development, need to incorporate climate risk scenarios, like extreme heat and increased flooding. AI digital twins can simulate multiple scenarios and allow underwriters to price these bonds more accurately (Tomorrow.City 2025) (Fig. 13.3).

13.3.3 AI and Automation of Compliance and Audit Systems

With the automation of sensing and observational data and the ability to continuously monitor projects and processes, properly designed systems will be able to use natural language processing (NLP) and LLMs. These tools integrate the rules, regulations and technical requirements of a specific carbon-reduction pathway and project design, and analyze the format and content of required reports to the relevant registry or governmental agency to meet compliance and audit requirements.

Application for Green Finance

AI applications can be designed to improve the integrity of carbon-reduction systems and help significantly scale up these systems. As discussed above, the use of AI systems in the process of verification of carbon credits makes possible near-real time, continuous digital monitoring of data sources; a significant improvement from the current practice of periodic, manual sampling and auditing of data.

A potential application in this regard would be creating generative AI co-pilots for auditors. Such a co-pilot could ingest all project data and any relevant regulatory and technical requirements. With this information in hand, the co-pilot could create a complete audit report and verification statement in compliance with the regime in question. With the use of such a co-pilot to automate much of the current auditing process, auditors and regulators could focus on those complex questions that require human judgment and help to significantly increase the speed and effectiveness of the auditing process.

AI models can allow city auditors and public regulators to continuously review public disclosures and other market information in order to detect anomalies that are relevant to bond eligibility and carbon credit generation. Integration of programs like Explainable AI (XAI) can help flag breaches in compliance, and deliver auditability for investors and regulators (ResearchGate 2025). Another potentially valuable application would be the use of AI models to produce dynamic, continuously adjusting baselines. In this case, the impact of an unexpected event (e.g., a large wildfire) on the project being monitored could be incorporated into the calculation of additionality of the project with a dynamic adjustment in the resulting issuances of credits.

13.4 Conclusion

13.4.1 Considerations When Using AI for Urban Green Development

Project developers, city leaders and climate investors continue to look for ways to scale up carbon reduction projects, increase liquidity of carbon credits and bolster confidence in carbon markets writ large. Based on recent insights regarding the useful deployments of AI technology, there are a number of important considerations that designers of green financial systems and carbon markets should bear in mind. Since many elements of any future digital infrastructure for AI-enabled green development have yet to be tested and deployed at scale, these considerations are both proactive and preventative. Because the actual dimensions of any built-out future infrastructure will likely look quite different than today's designers can imagine, taking a holistic view at every stage of implementing AI solutions increases the likelihood of productive and beneficial outcomes.

Notably, these design considerations emphasize the 'how' of AI development and deployment, in addition to technical considerations, since the process of innovation and creation itself will likely be increasingly important to determining the economic and environmental impacts of future economic development. Cities serious about incorporating these considerations will better prepare themselves to leverage AI in a way that unlocks the trillions of dollars of new green financial investment required to meet the world's climate development challenges.

13.4.2 Transparency is Essential

Declining levels of trust in the public institutions responsible for the development and oversight of economic activity impact the scale and rate of growth of green finance and carbon markets. Local governments facing ongoing budget constraints may require a higher level of confidence in the efficacy of new expenditures at a systems level. The desire for demonstrations of the efficacy and safety of new approaches and projects will likely be magnified with the use of AI and related digital technologies. Heightened concerns with the broader public about potential dangers of these technologies in general will likely make policymakers require more derisking of new technology deployments. Relatedly, continuing concerns about greenwashing in voluntary carbon markets and green development projects underscore the value of building in transparency with digital MRV in all projects, in real time, throughout their full lifespan.

13.4.3 Build AI Guardrails

Because AI-enabled development will increasingly be marked by its speed and scale, city stakeholders will need to seriously consider the potential long-term social, environmental and economic challenges of a deployment or application. From initial feasibility studies and development agreements to the governance of smart contracts for token issuances, the potential unanticipated consequences should be considered early in the planning process.

Powered by AI applications, the scale of damage from unconsidered 'externalities' could be larger and more difficult to address than with development to date. One current method to incorporate this approach would be to use explainable AI (XAI) in the project design and incorporate underlying data governance protocols in the first stages of development. Notably, the use of frameworks for innovation, like regulatory sandboxes, should not be counter posed with the creation of protections from negative economic and social impacts. Innovation with guardrails will be essential to creating projects that deliver economic, social and environmental value.

13.4.4 Plan for Continuous Change

To be successful in scaling low carbon urban development, it will be vital that designers, implementers and operators incorporate a learner's mentality as a core part of incorporating AI into their practice to enable them to adapt to the inevitable changes urban developments will face. Insights gained from early pilots and deployments indicate that many aspects of future AI-enabled systems will look very different from current applications. The technological breakthroughs of the 21st century, including the advent of mass adoption of artificial intelligence, have driven change in the world at a nearly unimaginable pace and scale.

References

Al-Fuqaha et al (2024) A hybrid CNN-LSTM-transformer model for IoT networks anomaly detection. IEEE Internet Things J 11(4):1025–1038. https://doi.org/10.1109/JIOT.2023.3325021

ATACH Community of Practice (2024) Bangalore's AI dengue monitoring platform. In: The Rockefeller Foundation Urban Climate-Health Action Report

Biljecki et al (2025) AI model maps building emissions to support fairer climate policies. Nature Sustainability. NUS College of Design and Engineering

Burns (2025) R-zero software designed to support LL97 compliance in NYC. Facilities Dive. https://www.facilitiesdive.com. Accessed 8 Dec 2025

Deloitte Canada (2023) Deloitte Canada acquires Fleet Challenge Canada Inc. to grow its Sustainability practice. Corporate Press Release

CMA (2025) The Green Bond Principles: Voluntary Process Guidelines for Issuing Green Bonds. International Capital Market Association, Zurich/London

De Wilde (2023) How Satellite Data Is Used to Detect Deforestation. Meteory Blog

Edinger (2025) San Jose is a 'Civic Sandbox' for Artificial Intelligence. Government Technology

Federation of Canadian Municipalities (FCM) (2023) Case study: Updating municipal vehicles to lower emissions—and costs. Green Municipal Fund

Feyisayo A et al (2024) AI-driven decarbonization of buildings: Leveraging predictive analytics and automation for sustainable energy management. World J Adv Res Rev 24(01):061–079

Le Fur (2025) Can AI strengthen ESG? Why traceability and ethical design matter. TRENDS Res Advis15(12)

Girgekar (2025) Leveraging AI for Resilient Urban Disaster Management in India. E3S Web Conf 632:02009

Gołasa P, Lizińska W, Wysokiński M (2025) The voluntary carbon market: market failures and policy implications. In: Environmental governance and economic sustainability. Springer, Cham. [*Note* Reflects recent 2024/2025 findings on standardization]

Guizar-Coutiño A, Robert CP, Coomes TRH et al (2020) Overstated carbon emission reductions from voluntary REDD+ projects in the Brazilian Amazon. Proc Natl Acad Sci (PNAS) 117(37):22802–22808. https://doi.org/10.1073/pnas.2004334117

International Capital Market Association (ICMA) (2025) The Green Bond Principles. Voluntary Process Guidelines for Issuing Green Bonds

International Energy Agency (2021) Empowering Cities for a Net Zero Future. IEA, Paris

IPCC (2022) Climate change 2022: mitigation of climate change. Contribution of working group III to the sixth assessment report of the intergovernmental panel on climate change, chapter 8. Cambridge University Press

Islam FAS (2025) Artificial intelligence-powered carbon market intelligence and blockchain-enabled governance for climate-responsive urban infrastructure in the global south. J Eng Res Rep 27(7):440–472. https://doi.org/10.9734/jerr/2025/v27i71585

Jia X, Bo Y (2025) Augmented finance for climate action: a systematic review. Sustain Financ Invest Rev 7(1):42–58

Kumar S, Lim WM, Sivarajah U, Kaur J (2023) Artificial intelligence and blockchain for transparent carbon markets: policy and technology gaps. Int J Inf Manag 72:102660. https://doi.org/10.1016/j.ijinfomgt.2023.102660
Mayors of Europe (2025) Barcelona's AI-Driven Mobility Transformation. Urban Innovation Reports
Naran B, Shankar V, Fernandes PdA, Dixon J, Burnett J, Abraham S, Stout S, Connolly J, Strinati C (2025) Global landscape of climate finance 2025. Climate Policy Initiative, London/San Francisco. https://www.climatepolicyinitiative.org/publication/global-landscape-of-climate-finance-2025/. Accessed 24 Dec 2025
OECD (2024) OECD survey on drivers of trust in public institutions—2024 results: building trust in a complex policy environment. OECD Publishing, Paris
Owens O (2025) Sustainable resource optimization using IoT and smart environmental monitors in urban infrastructure projects. Br J Earth Sci Res 13:1–17. https://doi.org/10.37745/bjesr.2013/vol13n3117.
Probst B, Schneider L, Cames M et al (2024) Systematic review of the actual climate impact of carbon offsets. Nat Commun 15:8432. https://doi.org/10.1038/s41467-024-53131-y
Smith JA, Zhang L (2025) AI-driven greenhouse gas monitoring: enhancing accuracy, efficiency, and real-time emissions tracking. AIMS Environ Sci 12(3):495–525. https://doi.org/10.3934/environsci.2025023
Soubry I, Alemu R (2024) Predicting future climate scenarios: a machine learning perspective on greenhouse gas emissions in agrifood systems. Front Environ Sci 12:1471599. https://doi.org/10.3389/fenvs.2024.1471599
S&P Global Ratings (2024) Sustainability insights: research suggests data gaps may be hiding greenwashing risks. S&P Global Ratings Research. https://www.spglobal.com/ratings/en/research/articles/240118-sustainability-insights-research-suggests-data-gaps-may-be-hiding-greenwashing-risks. Accessed 24 Dec 2025
S&P Global Ratings (2025). Global sustainable bond issuance to hold steady at $1 trillion in 2025. S&P Global Ratings Research. https://www.spglobal.com/ratings/en/research/sustainability-insights
Tomorrow.City (2025) Building climate resilience: how cities use digital twins. Published Nov 2025
United Nations Department of Economic and Social Affairs (UN DESA) (2024) World Population Prospects 2024: Summary of Results. Population Division
Venkateswarlu et al. (2025) Sustainable innovations in digital twin technology: a systematic review about energy efficiency and indoor environment quality in built environment. Front Built Environ 11. https://doi.org/10.3389/fbuil.2025.1523464
World Bank (2022) Digital monitoring, reporting, and verification systems and their application in future carbon markets. Washington, DC. License: CC BY 3.0 IGO
World Bank (2025) Urban development overview. World Bank Group
World Economic Forum (2018) The Oslo model: how to prepare your city for the electric-vehicle surge. WEF White Paper
Zhong S, Liu X (2025) Safety detection method for electric bike riding incorporating license plate region recognition. IEEE Access. https://doi.org/10.1109/ACCESS.2025.11079553

Part VI
Networking Initiatives on AI Applications for Climate Solutions at the City Scale

The UN Climate Change Technology Executive Committee (#AI4ClimateAction) is a key initiative connecting stakeholders on AI for climate action, particularly in developing countries, by identifying opportunities, addressing challenges like bias and inequity and fostering capacity building and innovation. This includes technical papers on AI's potential, a global forum for sharing best practices and the launch of innovation challenges and research networks to scale responsible AI solutions for mitigation and adaptation.

Key aspects and initiatives include the AI for Climate Action Forum whose activities are:

- Publication of technical papers and reports

•Innovation and challenges

- •Research and capacity building
- Focus on developing countries

•Addressing risks and challenges.

This part includes two chapters, showcasing applications of AI to revolutionize climate solutions in cities of different continents.

Chapter 14
Proposed AI-Driven Eco-Cultural Tourism in the Song Dam Lake and its Surrounding Area, Da Nang City, Vietnam: A Case Study of AI Project-Based Learning and the Impact of Eco-Cultural Tourism in Sustainable Development

Kwi-Gon Kim, Trieu Thy Hoa, Catherine Atkin, Anna Grichting, Ken Yocom, Seijoong Kwon, Hyunwoo Nam, Ho-Gil Park, Anton Bozhedarov, Nikolay Shvetsov, Bong-Koo Lim, Michael Schmitz, and Bui Quang Phong

Abstract AI eco-cultural tourism meets sustainability. There are many problems to be solved with the prediction and optimization modeling processes developed in this book, since we lack experience in using the process in the AI industry. The primary purpose of this chapter is to emphasize how new AI technologies can improve planning processes and outcomes and promote inclusive and sustainable developments

---

K.-G. Kim (✉)
Seoul National University, Seoul, South Korea
e-mail: kwigon@snu.ac.kr

T. T. Hoa
Quang Nam University, Da Nang, Vietnam
e-mail: tricuthyhoa@gmail.com

C. Atkin
Stanford Codex Climate Data Policy Initiative, San Jose, USA
e-mail: catherine.atkin@codex.stanford.edu

A. Grichting
University of Vermont, Burlington, USA
e-mail: bordermeetings@hotmail.com

K. Yocom
University of Washington, Seattle, USA
e-mail: kyocom@uw.edu

S. Kwon
Ministry of Foreign Affairs, Seoul, South Korea
e-mail: sjkwon@mofa.go.kr

K.-G. Kim (ed.), *AI Transformation of Cities for Climate Solutions*,
https://doi.org/10.1007/978-3-032-21940-4_14

by taking sustainable eco-cultural site development planning as a case study. More specifically, the chapter explores potential applications of AI for the ecosystem health of the selected lacustrine wetland, focusing on carbon sequestration and biodiversity. This chapter describes a test case for practical applications of the AI agents and work tools described in this book.

Keywords Project typology · AI-based zoning · AI-based eco-cultural tourism · Community development · Project work plan · Implementation strategy · Physical indicators · AI local participation agent

14.1 The Proposed Project Typology

A number of high-level conservation and development typologies have been formulated as a guide for the nature and character of zoning, conservation plans, urban design and eco-cultural tourism plans.

These typologies are mixed use, mimicking nature and reflecting international best practices for vibrant and sustainable AI-based eco-cultural tourism and community development to shape Southeast Asia's AI capital for the twenty-first century.

H. Nam
SimPlatform, Seoul, South Korea
e-mail: hwnam@simplatform.com

H.-G. Park
ArkLABS, Seoul, South Korea
e-mail: wonja720@daum.net

A. Bozhedarov · N. Shvetsov
AbaCapo, Berlin, Germany
e-mail: a.bozhedarov@avacapo.com

N. Shvetsov
e-mail: n.shvetsov@avacapo.com

B.-K. Lim
Asia Environment Institute, Seoul, South Korea
e-mail: ecotourism@naver.com

M. Schmitz
Stanford Codex Data Policy Initiative, San Jose, USA
e-mail: michael.schmitz@codex.stanford.edu

B. Q. Phong
Ban Thach Ward, Da Nang, Vietnam
e-mail: buiquangphong@gmai.com

14.2 Project Site Description

Photos 14.1, 14.2 and 14.3 show the international and regional locations, satellite imagery and scenes of the case study site of Song Dam Lake and its surrounding area in Vietnam.

(a) Flying storks (Photo 14.3a)
(b) Resting storks in their habitats (Photo 14.3b)

14.3 Objectives

- Overall objective: To develop a new AI-based planning and development process
- Specific objectives: To conduct the following six primary project activities

(1) AI-driven integrated water-quality monitoring system
(2) Creation of AI climate and eco-tourism agent
(3) Development of habitat-suitability index for storks: AI habitat agent
(4) Calculations of carbon sequestration potential: AI carbon Agent
(5) Engagement of stakeholders and technical experts: AI local participation agent
(6) AI eco-cultural tourism planning and management for natural climate solutions.

Figure 14.1 shows the AI optimization modeling objective function used to achieve the objectives described above.

14.4 Project Activities, Outputs and Outcomes

The proposal will connect the project activities to quantifiable outputs and ultimately desired outcomes (Table 14.1). This enables reviewers to evaluate what the proposed project would achieve, if implemented.

14.5 Multi-AI Agents for Activity Optimization

The multi-AI agent method described in this book will be tested and verified through the proposed case study (Fig. 14.2).

Photo 14.1 Location and satellite imagery of the selected case study area

Photo 14.2 Song Dam Lake wetlandscape. *Source* Binh (2023) Exploring the pristine beauty of Bai Say in the Song Dam area, Quang Nam (translated from Vietnamese). Dân trí (in Vietnamese)

14.6 Project Work Plan: Phased Roadmap

This is a phased-in project, and the activities for the four different phases are outlined as follows.

Phase 1

- Installation of sensors and development of local participation survey tools
- AI-driven data collection, baseline data analytics, prediction and optimization

(1) Development of AI agents
(2) Water quality and quantity
(3) Habitat-suitability evaluation
(4) Carbon emissions and sequestration
(5) Local participation
(6) AI eco-cultural tourism and its surrounding community development

Phase 2

(1) Preparation of land-use plan
(2) Preparation of master plan
(3) Stakeholder workshop on the draft integrated master plan
(4) Sector-specific plans
(5) Interim report

Phase 3

(1) Timeline
(2) Capacity-building plan

(a) Flying storks

(b) Resting storks in their habitats

Photo 14.3 Stork habitats. *Source* Binh (2023) Exploring the pristine beauty of Bai Say in the Song Dam area, Quang Nam (translated from Vietnamese). Dân trí (in Vietnamese)

(3) Investment plan
(4) Detailed budgeting plan

Phase 4

(1) Specifications for construction
(2) Cost estimate sheet
(3) Bidding (tender)

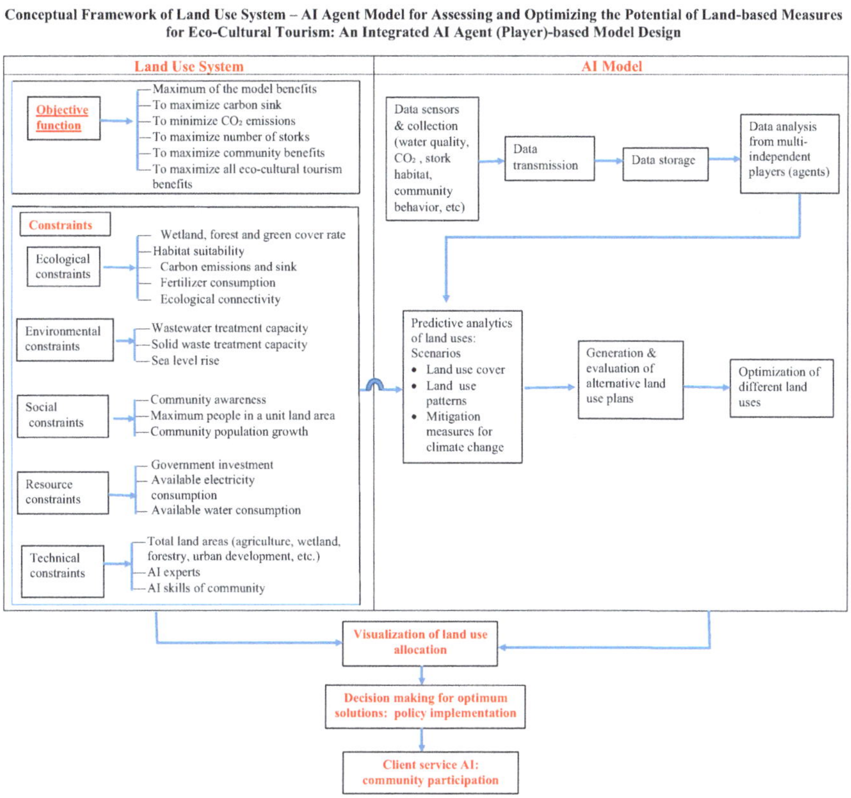

Fig. 14.1 AI optimization modeling process. *Source* authors

14.7 Implementation Strategy

14.7.1 Strategic Considerations

To develop initial recommendations in the advisory group meeting, a meeting with AI experts was first held in Seoul, Korea. The following considerations form the implementation of the proposed project, and integrate 'AI thinking' into future initiatives.

Summary of the Pre-meeting for the Preparation of the Project Implementation Plan

(1) Agenda

- Greetings

Table 14.1 Proposed project activities, and expected outputs and outcomes at the end of the project

Project activities	Outputs at the end of the project	Outcomes
1. Water-quality monitoring • Sensor installation • Monitoring activities • Data analytics • AI pollution control measures	12 sensors and monitoring stations installed and demonstrated in the site 60 smart water quality meters (physical water meters, AI digital communication components) installed 15 AI jobs created in the site 33 fish species in the lake 400–500 storks observed in the site	• Improved water quality • Improved community livelihood • Increased fisheries resources • Increased beautification • Increased numbers of tourists and scientific visitors
2. AI agent	4 AI agents 60 sensors 3 workstations 2 cloud servers 1 super AI computer 2 AI computer control systems 6 AI tools 5 AI programs and web algorithms 4 skilled AI experts 1 AI data center built	• Advanced and customized AI Agent ecosystem for wetland ecosystem conservation and development
3. Habitat suitability	A new AI agent 12 sample plots established 16 sensors and digital meters installed 20 acoustic sensors installed 3,500 mitigation measures suggested including 500 trees and 3.000 herbaceous plant species planted 7 measurement techniques used 18 researchers involved 4 monitoring programs structured	• Increased number of storks • Natural capital • Increased predictive ability • Wiser water-oriented development • Increased formation of stewardship network • Increased watershed protection • Improved health of the habitat

(continued)

Table 14.1 (continued)

Project activities	Outputs at the end of the project	Outcomes
4. Carbon sequestration	6 chlorophyll meters installed 10 soil sensors installed 180 tonnes of CO_2 stored in living vegetation 12 tonnes of CO_2 stored in litter 450 tonnes of CO_2 stored in organic soils and sediments 50 tonnes of carbon emissions reduced 4 vegetation sensors installed 60 near-infrared photographs which indicate decreased or increased photosynthesis 2 drones used for data or sample collection 3 water control structures (e.g. saline water intrusion, storm water, etc.) installed 2 AI smart water controllers installed 1 clean development mechanism (CDM) certified 10 CO_2 emission dashboards	• Increased carbon sequestration • Increased carbon responsiveness
5. Local participation	3 AI model local participation plans 100 local residents involved 100 million VND/1 year of local income generated 100 jobs created 3 community livelihood projects implemented 1 AI local eco-village demonstrated 2 AI local participation programs in operation	• Increased local awareness • Improved community livelihood

(continued)

Table 14.1 (continued)

Project activities	Outputs at the end of the project	Outcomes
6. AI eco-cultural tourism	3 AI eco-tourism models and tools developed 5 plans and maps prepared 2 demonstration tourism sites constructed 1 AI education, training and research center established 3 AI infrastructures installed 1 AI parking lot installed 6 physical infrastructures (roads, power supply, buildings, kiosks, etc.) installed 3 AI visitor, car and bike counts installed 4 tourist facilities developed 3 ecological restoration features 2 nature tech innovations applied 15,000 visitors expected 100 jobs created 150,000 dollars/year of tourism income created	• Increased sustainability of tourism • Increased non-consumptive benefits (aesthetic, spiritual, cultural/historical, existence value)

- Opening remarks (Hyunwoo Nam)
- Presentation (Kwi-Gon Kim)
- Open Discussion

(2) Date and Place

- Date: 14:00–16:30, September 25, 2025
- Place: Meeting room, SimPlatform, Seoul, Korea

(3) Participants

- International Advisory Group: Kwi-Gin Kim, Hyunwoo Nam
- SNU team: Prof. Bong-Goo Lim, President, Asia Environmental Research Institute
- SimPlatform team: Taehee Ahn, Board of Directors, AX Project Team, SimPlatform (NUBISON), six AI experts and scientists

(4) Main Points

The meeting raised the following considerations as the main points for success:

- In addition to the AI models and tools for the Song Dam Lake which are currently being developed, open-source AI models and tools may be

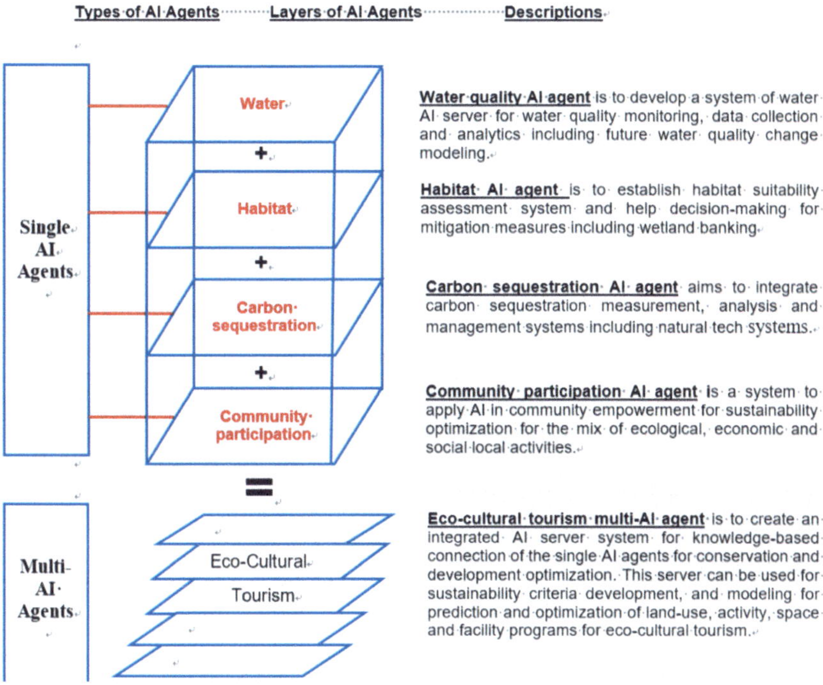

Fig. 14.2 Layers of AI agents. *Source* authors

customized for the Song Dam Lake project. The participants believe that the applications of AI models and tools proposed in the KOICA PCP are practically and technically feasible through a collaboration process.

- A project manager for each sector (water, climate, habitat, community participation, human behavior system and visual media art, eco-cultural site planning and design) may be needed to assist the project manager of the entire project.
- Computer-aided site base maps, including digital map data, should be available to show the deployment of different types of sensors on the actual site and on the proposed land-use map later. As soon as these data are available, the joint collaborative team should be ready to work on them.
- The operation and maintenance of the post-project period should be considered when the project work plan is designed.
- Making the International Advisory Group a formal part of the AI City Networking Initiative could help to streamline the implementation of the project.

(5) Photos

(a) Professor Kim's presentation on the Song Dam Lake project (Photo 14.4a)

(b) Participants in the discussion (Photo 14.4b).

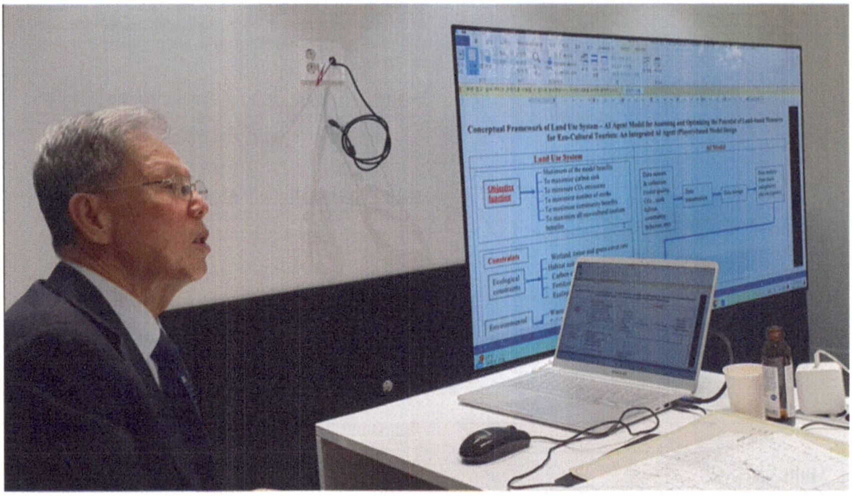

(a) Professor Kim's presentation on the Song Dam Lake project

(b) Participants in the discussion

Photo 14.4 The advisory group meeting

14.7.2 Implementation Strategy

An implementation plan will be prepared for each sector in conjunction with the work plan and implementation strategy.

Deployment of Sensors

The types of sensors and AI sensing techniques deployed for intelligent sensing for measurement and monitoring can be listed as follows:

- Water-quality measurement sensors
- Chlorophyll meters
- Photosynthesis measurement sensors
- Acoustic sensors
- Smart sensor in AI, working 24 h every day
- Real-time monitoring equipment (gas, water, electricity)
- Intelligent CCTV (AI CCTV)
- Location-data–based sensors.
- Solar-powered IoT (crop and soil) sensor to monitor plant photosynthesis
- Smart sensors to accurately measure and monitor carbon sequestration (carbon sink) and emissions including CO_2 sensors
- Harmful-gas-sensing and notification system
- Remote monitoring meters
- Soil-moisture monitoring sensors
- Vegetation-distribution (patterns) and health monitoring sensors
- Meteorological sensors (temperature, humidity, wind, etc.)
- Noise and fine-dust monitoring sensors
- IoT: Internet with 5G/5.5G/6G and WiFi in and around the tourism site.

14.7.3 Solutions by Sector

Water Quality

AI tools for water quality analytics include machine learning models like Random Forest, Support Vector Machine (SVM), Artificial Neural Networks (ANNs), and Long Short-Term Memory (LSTM) networks, which are used for predicting water quality indexes (WQI), classifying water quality, and forecasting future changes using historical and remote sensing data. Specific AI software and platforms use these models, analyzing data from various sources, including in-situ sensors, drones, and satellite imagery, to provide real-time monitoring and broad-area assessments.

In this case study, we use water quality monitoring, prediction and optimization of water quality changes with AI technology (e.g. NUBISON AIoT) in the Song Dam Lake and its surrounding communities.

A brief description of the five steps is as follows:

Step 1: Selection of water-quality indicators.

(1) Chemical indicators
   (a) Salinity
   (b) Hardness
   (c) pH
   (d) Acidity
   (e) Alkalinity
   (f) Contents of iron, manganese, chlorides, sulfates, sulfides, heavy metals, nitrogen and phosphorous

(2) Biological indicators
   (a) Biological oxygen demand (BOD)
   (b) Bacteriological parameters
   (c) Vegetation

(3) Physical indicators
   (a) Water depth or levels
   (b) Slope
   (c) Suspended solids
   (d) Sediment

Step 2: Deployment of sensors, IoT and AI equipment.

Step 3: Description of water-quality levels. The study uses sensors, documentary, library and field studies for data collection in the studied sites.

Step 4: Prediction and optimization of water-quality change, using AI for predicting and optimizing water-quality changes, including regression analysis.

Step 5: Pollution-control measures, using ecological restoration strategies.

Stork Habitat-Suitability Analysis

Table 14.2 shows the content of each of the project activities for a habitat-suitability analysis, and the proposed indicative budget associated with each activity.

Carbon Sequestration and Emissions

The main tasks in this sector include:

- Application of AI in the whole process of ecosystem modeling for the planner and environmental manager

Table 14.2 Activities and contents of AI-based habitat-suitability evaluation

Activity	Contents	Indicative budget (in USD)
Development of AI habitat suitability Agent framework	• AI modeling of habitat suitability • Defining of deep learning workflow	50.000
Selection of all the possible indicators	• Literature review • Expert knowledge • Brainstorming • Naver • Google	30,000
Data collection	• Habitat survey – Field survey form – Spot-checks – Remote sensing	60,000
Data analytics	• Preparation of habitat maps • Suitability, carrying capacity and risk analyses • Prediction and optimization for decision making	100, 000
Mitigation proposals	• Preparation of habitat conservation plans to protect storks and other species • Use of AI tools • Habitat control Structures • Mitigation banking	200,000

- Application of AI in ecological prediction and optimization modeling
- Application of AI deep learning workflow in the ecological modeling process for optimum solutions
- Prediction model of the carbon sequestration potential
- Optimization model of the carbon sequestration potential.

Local Participation Enhancing Community Value Through AI

Artificial Intelligence (AI) offers powerful tools to enhance and support community participation in the design and management of eco-tourism sites. There are key activities for integrating AI into the participatory models described previously. In this case study, Table 14.3 shows the use of an AI local participation agent to enhance community involvement, with the content of each of the project activities, and the proposed indicative budget associated with each activity.

Table 14.3 AI local participation agent

Activity	Content	Indicative budget (in USD)
Modeling of local participation plan	• Undertaking of local community inventory • Identification of various local people • Preparation of community maps • Understanding characteristics of local people	50,000
Data collection	• Use of AI-powered answer engine (Perplexity) – Local and indigenous knowledge • Community issue survey – Use of Chatbots	100,000
Data analytics	AI technics for prediction and optimization of community livelihood	100,000
Revitalization of existing local communities	• Promotion of cultural values and tourism • Improvement of physical, social and economic infrastructure • Development of AI infrastructure	100,000
Community participation in environmental decision making	Building of environmental stewardship and climate watchdog	50,000
AI capacity building, education & citizen participation in tourism and site management	• Use of AI video analytics and citizen apps • Meat and dairy emission dashboard • Learning of AI smart farming technology	100,000
Incorporation of project results	Feedback loop to AI eco-cultural tourism and other project activities from local participation	30,000

Eco-Cultural Tourism with Multi-AI Agents

AI tools offer diverse solutions for eco-tourism, including predictive analytics for managing tourist flows and resources, computer vision for biodiversity monitoring via drones and satellite imaging, and NLP-powered chatbots for multilingual traveler support. These tools also help minimize environmental and climate impact through optimizing visitor paths and hotel energy management, and they empower local communities by connecting travelers with local initiatives. Additionally, AI generates virtual eco-adventures and educational content to promote conservation awareness and minimize the need for physical travel to sensitive areas.

Listed below are some specific functions of AI tools and their applications in eco-tourism:

- Event portal for one-stop solutions, getting information on multiple events and providing feedback on those events through blogs, tweets, posts and other forms of social media
- Environmental and climate monitoring and management
- Resource optimization and sustainability
- Visitor experience and engagement
- Education and community empowerment
- Physical infrastructure.

In addition to implementation of sensors and solutions, the implementation should be viewed from AI physical infrastructure perspectives. Provision for AI infra should be part of the implementation plan.

The types of AI creative disruption technologies highlight AI-climate-sustainability nexus as basics to the proposed project need and possible AI solutions.

The case study intends to showcase: (1) the role of AI eco-cultural tourism in the area of Song Dam Lake and its surrounding communities, and (2) the methodology to promote urban sustainability by using the AI disruption technologies described in this book.

There are many advantages associated with AI applications in city climate and biodiversity planning and benefits accrued from the proposed project, including:

- Promotion of international networking initiatives
- Cooperation with UNFCCC[1] AI Climate Change Forum
- Enhanced brand value of Da Nang City, Song Dam Lake and its surrounding communities
- Development of various types of AI creative disruption technologies which can provide means for improving current practices of urban planning and management for climate solutions.

References

Binh C (2023) Exploring the pristine beauty of Bai Say in the Song Dam area, Quang Nam (translated from Vietnamese). Dân trí, 22 July 2023 (in Vietnamese). https://dantri.com.vn/du-lich/kham-pha-ve-dep-hoang-so-bai-say-song-dam-giua-long-quang-nam-20230722170948869.htm

Kim KG, Dang NBT, Trieu TH, Tran TH, Grichting A (2025) Ecologically determined contributions (LDCs) approach to AI wetland eco-tourism: the case of Song Dam Lake Wetland, Tam Ky City,

[1] UNFCCC stands for United Nations Framework Convention on Climate Change.

Vietnam. In LCA and consumption/needs-based GHG accounting for climate action: a pathway to carbon neutrality. Springer Nature, pp 231–263

Saint Mary's University of Minnesota, Geospatial Services (2013) A landscape-level approach to wetland functional assessment: Wetland mapping & functional assessment, Canadian River Watershed, New Mexico [Technical report]. Saint Mary's University of Minnesota

Chapter 15
Why African Cities Need to Adopt/ Integrate AI into Their Urban Planning Processes: A Dive into the Current Planning Challenges and Opportunities

Samuel Yawe

Abstract Most African cities are experiencing and grappling with numerous complex challenges of our time such as high population growth, rapid and uncontrolled urbanization, severe climate change shocks, infrastructure strains, natural resource management threats and housing deficits among others. These have caused a negative outlook on African cities and, hence, these cities have been characterized by traffic congestion, slum development, a brown agenda, encroachment on fragile eco-systems and increased climate shocks and risks such as floods. Therefore, the need for city/urban management has compelled local governments and city authorities to engage with city/urban planners through a robust planning process to find innovative ways of overcoming the city-related challenges so as to enhance city/urban performances. As part of the innovative efforts and approaches adopted in the planning processes, city planners have adopted a number of approaches ranging from 'ubiquitous city' (U-city concepts) to smart cities with the use of artificial intelligence (AI) as guiding principles in city planning. This chapter presents a comprehensive study of African cities and why they need to integrate AI into their planning processes, with emphasis on the current planning challenges and opportunities for African cities and urban planning processes in African cities, focusing on current realities. Additionally, an examination of artificial intelligence in the global urban planning context including global best practices and city case studies, lessons and opportunities for African cities, and reasons why African cities need to integrate AI into their planning process to meet the needs of the city dwellers as well as the trends of the current times.

Keywords Smart cities · Artificial intelligence · Urban planning · Climate change · Planning approaches · Concepts and processes · Urbanization · U-city concepts · Cities and climate risks

Y. Samuel (✉)
Rakai District Local Government, Kampala, Uganda
e-mail: yawehsamuel@gmail.com

K.-G. Kim (ed.), *AI Transformation of Cities for Climate Solutions*,
https://doi.org/10.1007/978-3-032-21940-4_15

15.1 Introduction

Most African cities are experiencing and grappling with numerous complex challenges of our time such as high population growth, rapid and uncontrolled urbanization, severe climate change shocks, infrastructure strains, natural resource management threats and housing deficits among others. These have caused a negative outlook on African cities and, hence, these cities have been characterized by traffic congestion, slum development, a brown agenda, encroachment on fragile eco-systems and increased climate shocks and risks such as floods. Therefore, the need for city/urban management has compelled local governments and city authorities to engage with city/urban planners through a robust planning process to find innovative ways of overcoming the city related challenges so as to enhance city/urban performances. As part of the innovative efforts and approaches adopted in the planning processes, city planners have adopted a number of approaches ranging from 'ubiquitous city' (U-city concepts) to smart cities with the use of Artificial Intelligence (AI) as guiding principles to city planning.

This chapter presents a comprehensive study of African cities and why they need to integrate AI into their planning processes, with emphasis on the current planning challenges and opportunities for African cities and urban planning processes in African cities, focusing on current realities. Additionally, an examination of artificial intelligence in the global urban planning context acknowledges global best practices and city case studies of AI integration in global smart cities (looking specifically at Korea). It offers lessons and opportunities for African cities and reasons why African cities need to integrate AI into their planning processes to meet the needs of their inhabitants, while also meeting the trends of the current times.

Also, this chapter will give a detailed look at AI applications for city planning while providing a systemic tour of African city planning approaches in regard to the current 4th industrial revolution (technological era), and make practical suggestions for city planning in terms of AI integration and usage as a key to solving city challenges and, hence, making African cities sustainable, smart, vibrant and livable.

15.2 African Cities

Africa is undergoing significant economic growth and associated demographic changes, including rising urbanization without the requisite infrastructure, spatial and settlement planning. The proportion of urban residents living in informal settlements is higher in Africa, as a region, than any other part of the world (UNDP 2012). Inhabitants of informal settlements often have inadequate access to basic utility services, lack security of tenure and experience precarious environmental, health and slum conditions.

The current economic growth and urbanization experienced in Africa needs to be guided by appropriate planning approaches, strategies and technology at all levels:

continental, regional, national, sub-national, and local. This will ensure that Africa's new towns, cities and rural areas are effective in terms of meeting the basic needs of all their inhabitants and that economic and social growth are inclusive.

Africa's urbanization patterns and the associated planning interventions can be traced back to the precolonial, colonial, post-colonial and modern periods. These historical influences reveal much about how current African cities have developed. Despite the fact that global cities are integrating technological advancements and innovations into their planning systems, approaches and strategies, such as AI and other related technologies, few African cities have adopted such technologies into their planning systems. Even the few African cities that have adopted the new technologies are facing significant implementation challenges due to a number of issues, such as big digital divides, unreliable internet access, and lack of expertise in the usage of the technologies, among others.

Figure 15.1 shows African cities located on a map of Africa, based on a regional clustering approach.

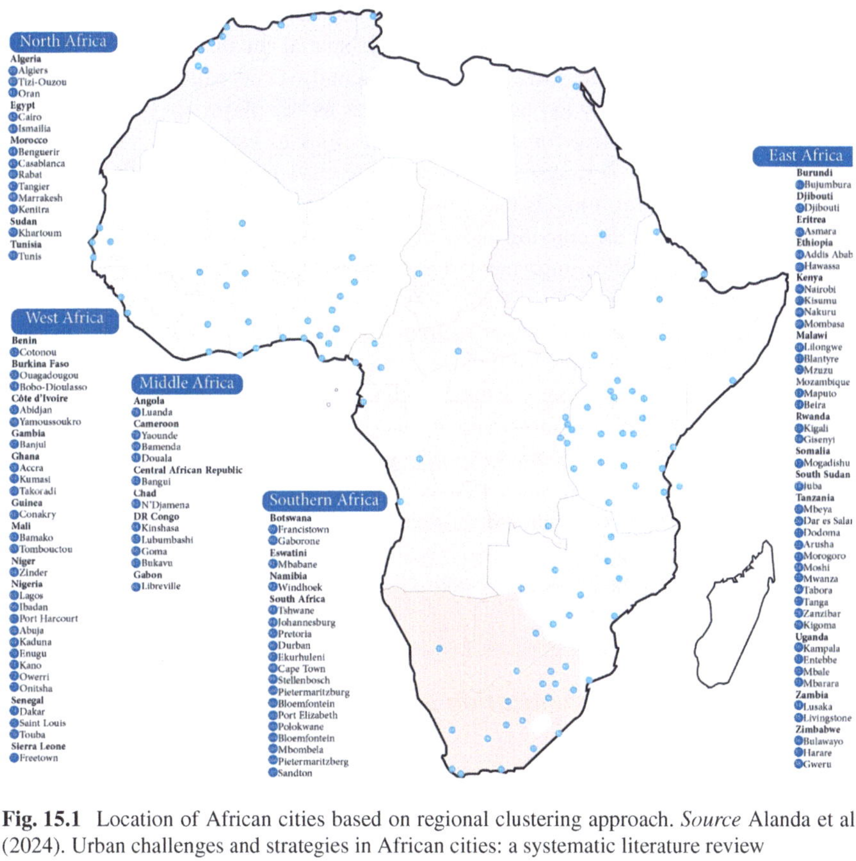

Fig. 15.1 Location of African cities based on regional clustering approach. *Source* Alanda et al. (2024). Urban challenges and strategies in African cities: a systematic literature review

15.3 Urban Planning in Africa

In discussing urban planning in Africa, one cannot fail to make critical observations on how African cities are planned. Two major observations that cannot be missed, were put forward by Bolay who said that it appears that the results of urban planning, as applied in developing countries, whether in Africa or elsewhere, only partially address the real problems facing urban populations. He also goes on to describe a caesura between planners, urban policy makers and residents, suggesting this is largely because the fundamentals of planning were built in the North and then transferred to the South, without having been redesigned on a clear basis, but simply adapted to other physical, political and financial constraints.

Watson (2009), while describing cities in the global South, said that demographic and territorial growth of cities in the South inevitably causes a concentration of poverty, social and economic, but also spatial inequalities in cities if urban planning is not practiced.

There is a need to entirely reconsider urban planning in African cities. This should begin from a participatory diagnosis in which the actual situation of the city is examined in its various dimensions, demographic, spatial and infrastructural, as well as considering the economic, social and environmental aspects which will give all the stakeholders an opportunity to position themselves as key players in city planning.

This makes it imperative for certain African cities to completely rethink their urban planning approaches, systems and strategies and remove from their current patterns of development, models that were designed for completely different environments, as a result of the copying and pasting of master plans with reproduction of defined standards and rules implemented in Western countries by specialists whose credentials are totally foreign to African countries.

This kind of planning projected African cities that would serve the interests of a minority group of citizens, favoring those groups who were at an advantage while leaving out the majority groups at the margins, who were living in informal and under-served neighborhoods, mostly the urban poor.

Therefore, rethinking the planning approaches presents an extraordinary opportunity to think about the future from what exists, taking into account the real resources, not only financial but also social, and to devise and implement urban planning with the goals of fighting against poverty and investing in equipment that has a sustainable impact on the living conditions of the poor.

15.3.1 Patterns of Urbanization and Associated Planning in Africa

Planning, and its associated patterns of urbanization and organization in Africa, can be traced back to the precolonial, colonial, post-colonial and modern periods. These have shaped the African cities we have today and have greatly affected current planning approaches and regulations. The following sections offer a brief look at these different periods.

Precolonial Period

Most parts of northern and central Africa were occupied by various ethnic groups and kingdoms during the precolonial period and were characterized by ongoing regional and continental migrations, occupations and the establishment of a wide range of small and large settlements. In West Africa, for example, this included the Mossie Kingdom which founded settlements such as Ouagadougou, Yatenga, Tengkodogo and Gourma (present-day Burkina Faso) and the Akan kingdoms, which included the inland Ashanti Empire (present-day Ghana).

Precolonial settlements in present-day Nigeria include Kano, Zaria, Koton-Karfi, Toro, Abeokuta and Ondo; Djenne and Timbuktu in Mali; Kilwa, Zanzibar and Dar es Salaam in Tanzania, amongst numerous others. In Southern Africa, many settlements (notably Great Zimbabwe, the capital of the Kingdom of Zimbabwe, and Bulawayo, Shaka's capital) were established by Bantu and Nguni peoples, settling in present-day Kenya, Zambia, Zimbabwe and South Africa, and joining the San, who were the first reported inhabitants of the southern region.

Many of these settlements accommodated large populations and were administered through traditional kingdoms. Land was vested within traditional leaders, families and communities, and controlled and managed through customary practices. Access to land was linked to usage and centered on agrarian practices, tribal villages and expansive settlements. In Nigeria, indigenous authorities administered settlements through spatial land allocation and management. For example, the Sokoto caliphate and large parts of the Oyo kingdom—seats of government—had a form of deliberate spatial arrangement of land uses around the palaces. By the middle of the 1800s many indigenous settlements, some with populations of up to 20,000, existed throughout the area, informed by the need for defense and religious practices. Settlement patterns and locations throughout Africa were also informed by the location of natural resources, land for grazing and other farming practices, watercourses and religious practices.

Colonial Political Systems and Planning

Modernist planning in Africa can be traced to its colonial roots and the interests of colonialists in the continent's natural resources, including agricultural produce, fresh water, minerals and, in particular, humans as slaves.

The colonialists were from countries that had relatively high levels of urbanization, and occupation was initiated through the establishment of new settlements along the African coastline from the sixteenth century onwards. It was common for settlers to establish new settlements away from existing traditional settlements, thereby limiting the growth of these settlements and ignoring the traditional land-management systems that existed.

Towards the close of the nineteenth century, European powers fervently vied for the acquisition of African territories; France imposed its governance upon West Africa, Britain imposed its governance on parts of West, Southern, and East Africa, while Portugal and Germany established their dominion in Southern Africa. In the wake of extensive colonial occupation, the planning systems that had been developed in Britain, Belgium, France, Portugal and other European nations were transposed onto the colonies. British town and country planning predominantly influenced East and Southern Africa (notably Kenya, South Africa, Zambia, and Zimbabwe) and certain regions of West Africa (including Ghana and Nigeria). Conversely, French planning was instituted in West and Central Africa (such as Benin, Burkina Faso, Chad, Côte d'Ivoire, Mali, and Senegal), while the Portuguese implemented their planning methodologies in Angola, Cape Verde, Guinea-Bissau, Mozambique, and São Tomé e Príncipe. These planning frameworks were primarily designed to regulate the progression of settlements, manage land use and oversee the construction of edifices within the colonies.

A pivotal objective of these planning systems was to facilitate the settlement of indigenous populations within colonial towns to satisfy the demand for inexpensive labor in the mining and manufacturing sectors. This was executed in a meticulously regulated manner, controlling the size and location of black townships under the pretext of health and sanitation concerns. Land use was circumscribed to residential housing and local retail establishments to safeguard the interests of colonial and white enterprises in the historic core areas. The concept of 'country' planning held limited relevance in rural locales, where land outside urban settlements was relegated to traditional authorities for management, supervised by a colonial administrator and, in certain instances, internal white authorities (e.g., South Africa, Zimbabwe). Colonial administrations endured for over a century in numerous countries. In contrast, the post-independence era has spanned merely a few decades, shrouded by the specter of dictatorships and military rule.

Post-Colonial Political Systems and Planning

Ghana was the first African country to receive its independence from the colonial powers (1957), while most of the other nations followed during the 1960s. Political instability was common for most countries throughout the initial independence period, characterized by coup d'états, political dictatorships, presidents who occupied office for life, the banning and imprisonment of political opposition and the induction of military governments. South Africa is notorious for institutionalized apartheid for close to half a century and the deliberate efforts to stifle the development of the majority of its population.

This pattern of political instability had a direct impact on the establishment of post-independence administrations and governance. In terms of development models, countries such as Tanzania (e.g., Ujama villages under President Julius

Nyerere, 1967) attempted to formulate indigenous practices, while others followed rural modernization practices as advocated by the colonial powers and aid agencies.

With the majority of the population rooted in rural areas, the emphasis was on rural planning as the spatial manifestation of development. In urban settlements, town and country planning frameworks were retained. In this way, the planning systems and laws introduced in the colonial era remained entrenched for the post-independence and current periods.

Modernist Planning

The proliferation of urban settlements in the post-independence era (from 1970 onwards) was influenced to varying extents—particularly in South Africa, which was significantly distorted by the apartheid ideology—by planning paradigms derived from the United Kingdom (such as garden cities and new towns like Milton Keynes) and the United States (such as the neighborhood concept, suburban developments and freeway expansion). Given that the administrations within the newly independent nations predominantly relied on an 'imported planning system,' the framework of planning administration remained largely unaltered.

Until recently, most African planners were educated at European and American universities and had no other knowledge base. Consequently, amendments to the British town and country planning system did not contest the foundational elements of blueprint planning, development control, permits and licenses. Development control, particularly in English-speaking African nations, continued to operate under Town and Country Ordinances, while proactive planning remained influenced by methodologies such as master planning, structure planning and spatial development frameworks. These planning paradigms were predicated on the assumption that future land uses could be precisely anticipated and orchestrated, and that all developments would predominantly be formal and modern.

This belief system stands in stark contrast to the realities of sluggish or stagnant growth in manufacturing and formal commercial enterprises, alongside the swift proliferation of informal settlements and slums in numerous African urban environments. Rural development has persisted as the purview of central government ministries and aid agencies, concentrating on development strategies characterized by significant state intervention, such as addressing basic needs, betterment schemes and integrated rural development programs.

15.3.2 *Why African Cities Need to Be Planned*

Intermediate cities are home to about 15% of Africa's population as these areas absorb most of Africa's urban population growth and it is expected that almost two-thirds of the people moving to urban areas will move to intermediate cities by 2040.

With an Urban growth rate of 3.5 annually for over the past 20 years, Africa's urban population is likely to double from 700 million people to 1.4 billion by 2025.

Therefore, using the 15% figure from the AfDB report which projects Africa's total population for 2026 at around 1.5 billions, with an urban population at around 50% by 2030,it is thus most likely to be from 47–48% in 2026 with a population of 705 to 720 million urban Africans in 2026. Hence, 15% of the total population would put intermediate cities at around 225 million people in 2026. It should be noted that the AfDB explicitly says that the 15% refers to the total population but not just urban population hence 225 million becomes a better estimate for 2026.

Consequently, urban planning, development, and redevelopment initiatives must be concentrated in these cities that have transitioned from villages and small towns, particularly in peri-urban regions. Numerous rural settlements have been elevated into administrative and local government hubs through direct political intervention, subsequently annexed into existing urban areas.

Current trends reveal a disproportionate emphasis on primate cities, typically capitals, and strategically located urban centers drawing substantial infrastructure and resource investments. In contrast, many secondary cities, smaller towns and rural locales suffer from a dearth of essential services and social amenities (Kadiri 2012). The burgeoning urban populace is characterized by a disproportionately high proportion of youth, elevated rates of unemployment and underemployment and a significant reliance on the informal economy. The percentage of urban residents residing in informal settlements is higher in Africa, as a region, than in any other part of the world (UNDP 2012). Informal settlements, often lacking secure tenure and existing in environmentally precarious conditions, house the majority of inhabitants in numerous African cities.

15.4 Artificial Intelligence

Artificial intelligence, as a technological domain, is perpetually evolving, rendering it challenging to encapsulate AI in its entirety. This relentless advancement has engendered a plethora of definitions of AI. AI has garnered substantial attention in recent years, yet it is far from a novel concept, tracing its lineage back to the advent of computers post-Second World War. The Dartmouth Conference in 1956 convened researchers from diverse disciplines to investigate 'thinking machines,' marking what is widely regarded as the inception of AI as a distinct academic field and the first recorded usage of the term 'artificial intelligence' by the pioneers at that conference.

There exists no singular, universally acknowledged definition of artificial intelligence; however, the Oxford English Dictionary characterizes AI as 'the capacity of computers, or other machines, to exhibit intelligent behaviour.' This implies that AI systems appear to think, learn, and act in ways akin to humans, and in certain instances, surpass human capabilities. AI systems possess the ability to analyze extensive datasets, resolve intricate problems, make informed decisions and undertake creative endeavors.

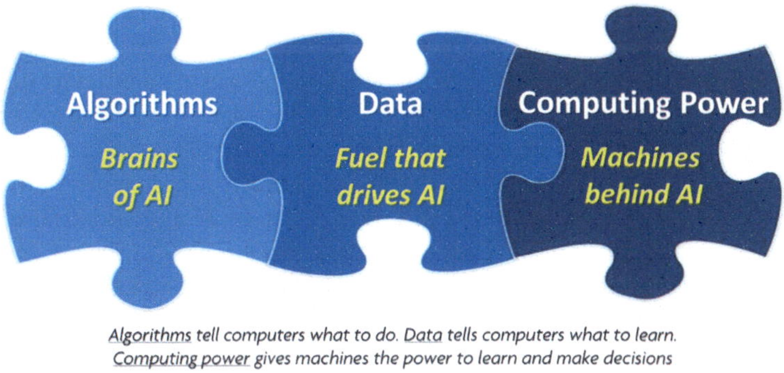

Fig. 15.2 Core elements of AI. *Source* World Tourism and Travel Council

In colloquial usage, AI often serves as an umbrella term encompassing a myriad of distinct types and technical subcategories. Fundamentally, AI is a system that generates outcomes based on a predefined objective.

The objective for AI entails the translation of a human-defined goal into a mathematical framework. Outcomes may manifest as predictions, recommendations or decisions. For instance, the human aspiration of triumphing in a chess match can be transmuted into the objective of selecting a sequence of moves that optimizes the likelihood of victory.

The terminology of AI algorithm, AI system, AI ecosystem and AI denotes varying scales of complexity. An algorithm represents the most specific construct; it is a process that yields an output from a given input. An AI system typically refers to a singular application that utilizes a unique input to produce its own output. An AI ecosystem denotes a network of AIs that engage in interaction with one another. AI, in its broadest sense, encompasses all the diverse methodologies and systems, usually referring to the field in its entirety.

Recent breakthroughs in AI can be attributed to three pivotal components: advanced algorithms, vast datasets and computational power that elucidate how machines are capable of exhibiting intelligent behavior and account for the sudden proliferation of AI into our daily lives. Figure 15.2 illustrates the core elements of artificial intelligence.

15.4.1 How Artificial Intelligence Has Been Applied in the World Today

Artificial intelligence has evolved into one of the most pivotal technologies in the modern world, profoundly influencing nearly every domain of human endeavor. Its inception can be traced back to the 1950s, and is attributed to John McCarthy, often hailed as the father of AI. Over the decades, it has played an instrumental role in advancing humanity, enhancing our capabilities to unprecedented levels. Notably, its applications have traversed a remarkable spectrum, extending 'from the soil to space.' The ubiquitous availability of the internet, the advent of cloud technology, the emergence of big data, sensors, and other technological innovations have collectively served as a boon for the advancement of AI.

Although AI has not yet supplanted humans entirely, it has significantly aided individuals in addressing complex and perilous challenges. Manifesting as intelligent machines, AI often acts as a surrogate for human beings.

In contemporary society, AI has permeated various sectors, including agriculture, business, education, entertainment, healthcare, defense and aerospace technology, where it has produced critical and commendable effects. In the healthcare sector, for instance, AI has played a transformative role in bridging the chasm between technology and medicine, particularly in patient diagnosis. This application necessitates data accuracy, security, and, crucially, the trust of patients in the system. Moreover, AI boasts remarkable applications in robotics, exemplified by robots performing surgeries or assisting in diagnostic processes. Fundamentally, AI manages vast quantities of data to derive the desired outcomes, with algorithms meticulously designed to mitigate errors and ensure precision. As previously noted, the breadth of its applications has indeed expanded from the soil to space, representing a burgeoning technology dedicated to the enhancement and advancement of human life.

15.4.2 AI Application and Use in Africa

Africa is increasingly adopting AI to unlock opportunities and address challenges in critical areas such as agriculture, health, legal services and the creative industries.

In agriculture, AI-powered technologies such as drone monitoring and predictive analysis are being harnessed across Africa to improve crop productivity for smallholder farmers in countries like Kenya and Rwanda to enhance food security.

Similarly, AI is transforming healthcare through early disease detection and supporting health data analysis, contributing to improved outcomes across Africa. The legal sector is also leveraging AI to expand access to justice, while the creative industries utilize AI to foster innovation and cultural expressions reflecting Africa's rich diversity.

15.4.3 AI Applications for Urban Planning

The term artificial intelligence (AI) encompasses a diverse array of methodologies that enable machines to exhibit intelligent behavior. In India, the rapid pace of urbanization presents significant challenges and opportunities for equitable and sustainable development. Consequently, the pursuit of innovative solutions for urban and regional planning has become an imperative necessity. In this context, AI is swiftly emerging as a trans-formative and promising technology poised to tackle intricate urban dilemmas and enhance the efficiency of planning processes (Batty et al. 2012).

Several AI technologies and algorithms that may prove advantageous in the realm of urban and regional planning in the future. It is important to note that the following list is not exhaustive, but rather provides insight into the current technologies in use.

Machine learning: This includes algorithms like neural networks supporting vector machines that help planners make predictions and identify patterns in complex situations and scenarios. It helps planners to synthesize complex datasets, discover hidden relationships between them, and forecast future patterns hence affecting their strategic choices or making decisions (He et al. 2020).

Geospatial analysis: Many professionals engaged in spatial planning practices extensively utilize geospatial technologies such as Geographic Information Systems (GIS), remote sensing and spatial econometrics for sophisticated spatial analysis and visualization. These AI-enhanced geospatial tools empower planners to amalgamate diverse spatial datasets across multiple dimensions and perform spatial interpolation and modeling. This capability enables them to discern spatial patterns and identify hotspots, thereby informing where interventions are necessary and delineating the nature of such interventions. AI-driven geospatial tools facilitate the integration of multidimensional data layers to recommend interventions based on prioritization.

Natural Language Processing: NLP is useful for analyzing vast amounts of textual data like plans, policies and public feedback and helps extract valuable insights from these texts. As an example, NLP can summarize qualitative information from large text bodies that reveal what different groups need and emerging issues in discussions.

Deep Learning for Urban Image Analysis: Deep learning is being examined to analyze urban pictures concerning infrastructure, building density and environmental factors. Deep learning is a branch of machine learning that draws inspiration from the structure and operations of the human brain, and machine learning enables computers to learn from data without the need for explicit programming. Once programmed correctly, the model can continuously analyze the imagery to check for changes and develop patterns.

Agent-Based Modeling and Urban Simulation: There is an increasing uptake of agent-based modeling supported by AI for simulating complex urban processes and evaluating different planning interventions that could assist stakeholders achieve their goals.

15.5 Global Planning Approaches

Global urbanization trends are increasingly influencing economic, social and environmental conditions on a global scale. Driven by rural-to-urban migration and escalating population growth rates, projections indicate that by 2030, nearly 60 percent of the world's populace will reside in urban centers.

Moreover, while cities occupy a mere three percent of the Earth's landmass, they are responsible for up to 80 percent of global energy consumption and 75 percent of worldwide carbon emissions. Confronted with challenges such as overcrowding, pollution, inadequate infrastructure, resource scarcity, health risks and crime associated with urbanization, governments are actively exploring alternatives to address the needs of their citizens without jeopardizing the welfare of future generations.

It was within this context, during the mid-1980s, that the concept of 'smart cities' emerged. This paradigm promised to usher in urban agglomerations that are sustainable, inclusive, interconnected, reliable and secure, while fostering a data ecosystem predicated on technological integration. Acknowledging the significance of intersectoral collaboration among diverse stakeholders—including governments, citizens, academia and the private sector—in the transformation of urban spaces towards smartness, it is imperative that civil society remains the fulcrum for the formulation and adoption of digital transformation solutions throughout the planning and operational phases of smart sustainable cities.

By harnessing technologies and applications across various sectors, cities can leverage the data streams generated to monitor, evaluate and enhance their operational performance, mitigate their environmental impact and cultivate greater sustainability. Consequently, the two principles of 'smartness' and 'sustainability,' once regarded as mutually exclusive, are now embraced as integral components of a comprehensive approach to managing urbanization across sectors, including transportation, education, energy, environment, waste management and healthcare, through the implementation of digital technologies.

Urban planners have long sought to integrate digital technology into their practices. Today, the pressures for greater efficiency in planning, design, and delivery are more pronounced than ever. A confluence of factors, ranging from escalating urbanization, challenges in public sector funding and intensified competition, compels design practitioners to achieve more with fewer resources.

In this context, artificial intelligence emerges as a formidable ally. The industry is increasingly turning to generative AI models to empower planners and designers in addressing urban challenges and conceptualizing solutions with enhanced speed and efficacy. The amalgamation of these generative AI models with the growing availability of real-time data enables practitioners to uncover innovative insights that are deeply rooted in the local context.

Imagine a city where AI is fully integrated into the design life cycle. AI-enhanced digital twins enable planners to anticipate and respond to climate events and to demographic shifts. Interconnected infrastructure systems create more responsive and integrated cities. AI breaks down silos between building services and allows structures to adapt in real time to environmental conditions and user preferences. Planning processes are fast, and, thanks to generative AI, citizens can see what their place will look like as it is designed, giving them the opportunity to be involved meaningfully.

The Korean government sees smart cities as an essential component for improving the quality of citizens' lives and as a means to grow Korea's economy through the development and commercialization of emerging 4th industrial revolution (4IR) technologies. Significant political support has been given to national smart-city projects, which are now scheduled for completion by 2023.

Smart cities are a strategic component of the Korean government's national economic policy for the fourth industrial revolution. The Korean government is seeking to grow Korea's economic competitiveness by taking a leading role in innovative technologies such as big data, artificial intelligence (AI), the Internet of Things (IoT) and blockchain. Smart city projects offer a way for the Korean government to develop its strengths in these areas through the development and commercial deployment of these technologies, much as it did with the core ICT infrastructure in the late 1990s. The government forecasts a total of GBP 68bn in economic effects from smart-city projects by 2030 out of a total of GBP 422bn from the fourth industrial revolution as a whole.

Korean smart cities have their roots in the U-city (ubiquitous city) of the 2000s. The U-city was Korea's first attempt at the integration of different services in a city through ICT, with the benefits of greater convenience, better quality of life and savings through the integration of different systems. Early examples of services included real-time information on bus and train arrival for passengers, integrated traffic management, including the enforcement of traffic violations, as well as the ability to access government services online (Fig. 15.3).

The development of Korea's smart cities is shown in Table 15.1.

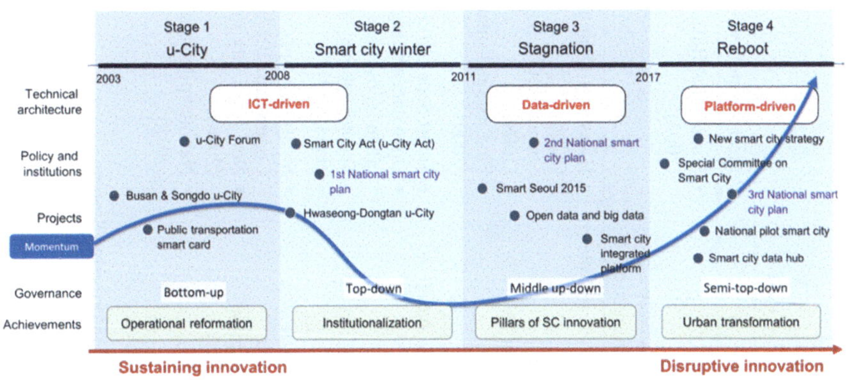

Fig. 15.3 The trajectory of Korea's smart-city development. *Source* World Bank compilation (2018). *Note* SC = smart city. In this figure, an 'ICT-driven' smart city is one that applies ICT-based systems, including urban informatics and digital applications; a 'data-driven' smart city actively utilizes data resources and analytics for city innovations; and a 'platform-driven' smart city applies platform-based ICT architecture that enables utilization of different platforms and systems, which are interoperable and which function based on shared city-level platform architecture

Table 15.1 Development of Korea's smart cities

Period	Characteristics	Project and policy	Approach	Key facts
Stage1: 2003–2007	*Start of U-city*: • ICT-driven • Sensor-based service • Operational reformation	• Songdo and Busan U-city (2003)	• Bottom-up approach • Development led by municipalities and urban developers	• Success cases in urban services innovations • City as a product approach, leading to slump of U-city
Stage2: 2008–2010	*Downturn of smart city (smart city winter)*: • ICT-driven • Institutionalization	• Hwaseong-Dongtan U-city (2008) • Special Act on Smart City (2008) • First national smart-city plan (2009)	• Top-down approach • Institutionalization led by national government	• National policy framework • Absence of smart city in urban development • Smart -ity plans, but lack of action
Stage3: 2011–2016	*Beginning of data-based innovations*: • Data driven • Pillars of smart city innovation	• Smart Seoul 2015 (2011) • Open data and big data for urban innovations (2012) • Second national smart-city plan (2013)	• Middle-up-down approach • Innovations led by municipalities • Institutions developed by national government	• Urban innovations not under the name of smart city • Beginning of disruptive innovations
Stage4: 2017–present	*Rebooting of smart city*: • Platform driven • City transformation	• Smart City Special Committee (2017) • New smart-city strategy (2018) • National pilot smart city (2018)	• Semi-top-down approach • Smart city strategy redesigned by national government • Municipality-level master planning and pilot projects	• Smart cities for the fourth industrial revolution • New smart-city concept and pilot project

Source World Bank compilation
Note TOPIS = Transportation Operation Information System

15.6 Why African Cities Should Integrate AI in the Planning Process: Challenges, Lessons and Opportunities

As global urbanization rapidly intensifies, cities are grappling with complex challenges, such as population growth, infrastructure strain and resource management. The inevitably increasing complexities of urban management have compelled local governments to seek innovative technologies to enhance urban and regional performance. As part of this effort, governments and related stakeholders have adopted the concept of the 'smart city' as a guiding principle for developing more efficient and suitable urban planning policies and practices.

However, a significant challenge remains: how best to effectively contextualize this global concept, which has yet to be fully embraced by African cities that aspire to be recognized as 'smart.' In the African context, where infrastructure gaps have hindered the region's pursuit of economic transformation, artificial intelligence (AI)-driven smart cities hold the potential to revolutionize public services, urban mobility, energy management and public safety.

As emphasized in Agenda 2063's Sustainable Urbanization for Africa's Transformation, harnessing technology and innovation is crucial for developing smart cities that enhance governance and service delivery and support communities and quality of life while promoting sustainability, efficiency and transparency.

As such, there is a need to scrutinize the potential of AI to address Africa's urban challenges. This will involve exploring how AI can enhance public infrastructure, bridge the digital divide and create a human-centered focus on inclusivity to ensure that no one is left behind in the continent's smart-city revolution.

15.6.1 Challenges Faced by African Cities

African cities are currently facing many urban challenges which are making them unpleasant places for the inhabitants. These challenges are human induced and also climate-change related and can largely be attributed to poor city planning and administration. In the search for city solutions, people like Myers (2010) suggest that African cities need to revise their urban theories and practices and ensure that they consider African perspectives instead of conceptualizing Western types of planning via a top-down paradigm.

He further suggests that African cities should consider African dynamics and should avoid the 'copy and paste game', especially in this era of rapid urbanization on the continent, as well as global climate change and environment degradation that require the consideration and integration of the local environmental and socioeconomic context. Table 15.2 shows a summary of the most pressing urban challenges faced by African cities.

Table 15.2 A summary of the most pressing urban challenges faced by African cities

No	Challenges/vulnerability	Cluster
1	Rapid population growth/urban migration	Spatial planning-governance (SPG)
2	Poor waste management/poor sanitation	
3	Congestion/traffic jams	
4	Insecurity/violence/crime	
5	City center decline	
6	Inadequate urban infrastructure/services/utilities	
7	Rapid urbanization/sprawl/urban expansion	Environmental-climate change (ECC)
8	Floods/landslides	
9	Environmental degradation/pollution/encroachment	
10	Land disputes/non-standard plot sizes	
11	Drought/water scarcity	
12	Displacements/exclusion/evictions	Political-socio-economic (PSE)
13	Slums/informal settlements	
14	Unemployment/poverty	
15	Food insecurity/urban hunger	
16	Socio-spatial inequality/urban injustice	
17	Unaffordable housing/urban inaccessibility	

15.6.2 Lessons and Opportunities for African Cities

Global cities that have adopted and implemented smart-city projects with an integration of AI into their city planning and they have provided valuable lessons that other cities of the world, including African cities, can adopt and learn from to improve their cities. Several Korean smart-city projects provide valuable insights and practical examples that underscore the reasons why African cities need to integrate AI into their city planning processes. Also, from a city planner's perspective, these international initiatives have influenced a new planning paradigm, shifting from the business-as-usual planning process to a new one that is smart, climate sensitive and technological, with a clear focus on the use of AI to solve, manage, predict and control city challenges and problems.

Therefore, this section highlights key lessons and opportunities for African cities in relation to the integration of AI into their planning systems and processes for better city planning and management (Table 15.3).

Table 15.3 Key lessons and opportunities for African cities related to the integration of AI into planning systems

Lessons	Opportunities
Set the vision and define a proper approach Global cities that have adopted AI for city planning have devised means to tackle a number of challenges and have shared their experiences. For example, the Korean government had a number of smart-city projects for cities such as Pangyo under the 'U-city' concept The vision for these city projects in the early development stage saw cities as a product and thus focused on constructing advanced urban infrastructure that did not achieve the goal as it offered limited opportunities and excluded collaboration, participation and innovation The Korean government later changed its approach and began to perceive cities not as products but as platforms that could provide common technical, physical and institutional grounds for a variety of functions and services. This later minimized the risks and costs of smart-city development, while encouraging innovation and collaboration within services and sectors Therefore, setting a vision and a proper approach for city development is crucial for a successful city planning initiatives	*Big data analysis capabilities* Big data provides opportunities to analyze very large and complex data which would be difficult to analyze by conventional data-processing applications. This makes it easier to analyze and process collected data for planning purposes, as big data provides insights that can improve decision making in the planning process. Such data includes, for example, information from sensors, social media and smart phones
Apply different strategies for different cities (be flexible and adaptable) Initially, fears, speculation and misconceptions arose that led to smart cities being seen as an advanced model of city development that could only be implemented in the developed world/cities. Yet smart cities simply represent a new approach; a strategy that offers an innovative method of urban development and is not a measure of a city's development rank. It is a new strategy that applies different technologies to inform decisions and manage cities so that they achieve their visions and goals while solving specific city problems and challenges. Global cities that have implemented smart-city projects have shown that every city can implement its own smart-city model by adopting different strategies and technologies to suit its goals and visions. Countries like Korea have also demonstrated that cities everywhere can adopt the approach and develop their own smart cities. This is demonstrated by their K-City Network Program, an exportable model for smart cities, funding almost GBP 500m worth of projects overseas	*Urban simulation* Urban simulations allow city planners to predict, anticipate and test the outcomes of different planning interventions before implementation, reducing uncertainty and costly mistakes. Advances in data availability, computational power and analytical methods now allows planners to model complex relationships between multiple urban factors such as land use, transport, population and environment The growth of computer-based visualization and simulation platforms, ranging from integrated urban systems to specialized prediction tools, strengthens evidence-based planning, scenario testing and long-term strategic decision making. Thus, simulations improve policy evaluation, support sustainable city planning and enhance the ability to present planning choices clearly to decision makers and the general public

(continued)

Table 15.3 (continued)

Lessons	Opportunities
Build a strong legal and institutional foundation Global cities investing in smart-city projects have clearly established that a strong legal and institutional foundation is key to a successful planning initiative. Strong governmental regulations help to avoid conflicts of interest between public duties and private interests. From Korea's U-city experience, we learn that establishing legal foundations and reinforcing organizational institutions and other relevant policies and strategies at an early stage of the smart-city development facilitates the creation of smart cities that are more sustainable, integrated and innovative	*Access to real-time data and short-term planning frameworks* Over time urban and regional planning timeframes have related to years—five years for short-term plans; 20 years and more for long-term plans. However, smart cities and big data are changing this approach by enabling planners to understand how cities function in real time. Cities can now be planned and adjusted over minutes and days, not decades. Real-time big data collected continuously and at every time interval allows urban dynamics to be monitored more frequently than traditional methods that could take ten years. This is a big opportunity for African cities to solve city problems and challenges in real time, as certain city problems require immediate planning response and action
Continue innovations through stakeholder engagement Developing smart cities is an ongoing process that involves continuous transformation. The real value of smart cities arises through the cultural transformation achieved when multiple stakeholders, including citizens, municipalities, and industries, act together. Citizens play a critical role as final users and co-developers of smart cities In the U-city era, citizens remained passive users, and the mismatch between the expectations of citizens and the initial goals of the U-city projects led to a downturn in smart cities. It is important to motivate and encourage citizens to engage in smart-city development so that they contribute to further innovations. The government and civil society should constantly increase awareness of smart cities' impacts through service provision and exercise leadership both in designing smart-city models and policies and in coordinating stakeholders. In Korea's experience, intergovernmental engagement across ministries and organizations strengthened smart-city leadership, while the broad participation of private sector players and start-ups spurred innovations	*Smart-city approaches* Global smart-city approaches shifted their focus and embraced a combination of top-down and bottom-up processes in their city planning processes. They emphasized the central role of smart citizens in shaping city development and urban spaces, which creates an inclusive public participation environment. These approaches have promoted open-access, open-source platforms and inclusive digital infrastructure, African cities can adopt these strategies to empower community groups, local associations and other stakeholders, as well as civic innovators to contribute local knowledge and insights into the planning process. This participatory approach helps to strengthen democratic urban governance, creates and improves plan legitimacy and creates a sense of ownership, as well as leading to context-sensitive solutions which enables African cities to develop smart-city strategies that are socially inclusive, locally grounded, resilient, vibrant and sustainable
Strengthen enablers of smart-city innovations Korea's experiences show that the role of the government in strengthening enablers of smart-city innovations has been critical. Smart-city enablers make significant differences in many aspects, including investment, collaboration, sharing of data and other resources and quality of services	*Data availability from crowd sourcing and digital platforms* In this era of technological advancement, cities already possess widespread and cost-effective sensor networks through citizen's smartphones that are equipped with sensors and applications such as GPS and accelerometers among others. These provide opportunities to collect real-time, high-resolution urban data without necessarily investing in expensive new infrastructure. Using smartphone-generated data collaboratively helps planners better understand their cities, such as mobility patterns, environmental conditions and public space usage. This enables evidence-based, responsive and inclusive planning as it minimizes costs, improves data availability and strengthens citizen participation in the planning process

(continued)

Table 15.3 (continued)

Lessons	Opportunities
	Built Competence and Confidence Most smart cities in the world have fostered competence and boosted confidence among city dwellers and travelers with the kind of information shared for public consumption. In return, citizens have shared trusted information and data that has enhanced better planning interventions. This is possible because there is increased usage of smartphones to access smart-city applications where these digital platforms have supported real-time sharing on transport, traffic, pollution, waste and other public services, enabling citizens to make informed and sustainable urban choices. For planning to be effective in African cities, planners need to learn from global smart cities and ensure that digital tools are user-friendly, simple, inclusive and accessible to even non-technical users so that citizens become more active contributors to the achievement of city visions, targets and goals
	Adopting the new urbanism approach New urbanism offers an important approach that focuses on promoting vibrant, walkable, compact, mixed use and well-integrated communities that ensure services, jobs, housing and public spaces are within close proximity. This planning approach directly responds appropriately to common challenges faced by African cities, such as urban sprawl, high transport costs, traffic congestion, environmental degradation and inefficient land use Therefore, the new urbanism approach provides African cities with a practical framework for creating livable, inclusive, compact, vibrant and resilient cities that improve the quality of life and long-term urban sustainability
	Sustainable urban planning frameworks and approaches Global smart cities have developed frameworks such as circles of sustainability that provide planners with a holistic tool/guide to assess, understand and manage urban developments across social, environmental, economic and governance dimensions, instead of focusing on just the spatial/physical growth of cities. Therefore, such approaches and frameworks can be adopted and customized by African cities to balance their development needs with long-term resilience, while ensuring that city-development projects address social inclusion, economic viability and environmental protection simultaneously. Thus, applying such sustainability frameworks helps planners to better prioritize investments, evaluate city development projects and guide cities towards resilient, equitable, vibrant, social and sustainable urban futures

(continued)

Table 15.3 (continued)

Lessons	Opportunities
	Availability of participatory tools and platforms Successful smart cities highlight the importance of citizen participation in the planning process. While participation takes many forms ranging from collaborative, deliberative or community planning, these cities have noted that its true value is embedded in moving from one-way information sharing towards a continuous dialogue, shared decision making and evaluation of outcomes Therefore, technological advancement has provided powerful participation tools such as social media platforms, digital feedback systems and other online engagement platforms which have helped to broaden citizen participation. These tools enable planners to engage with communities in planning, designing, budgeting and policy processes more effectively. This gives African cities an opportunity to strengthen planning legitimacy, improve responsiveness and ensure that urban interventions are more inclusive, people centered, context sensitive and socially and environmentally sustainable
	Availability of Sustainable Development Goal 11 and the New Urban Agenda The availability of SDG 11 and the New Urban Agenda is a key opportunity for African cities to reorganize and set new targets, goals and visions in the same direction as SDG 11 and the New urban agenda. SDG 11 states what cities must achieve by being inclusive, safe, resilient and sustainable, while the New Urban Agenda suggests how to achieve these goals through integrated city planning and management approaches. This framework will encourage African cities to adopt innovative planning strategies that bring a balance in social, economic, environmental and spatial sustainability. Hence, aligning city plans with SDG 11 and the New Urban Agenda helps to strengthen policy coherence, mobilize investments and guide African cities towards smart, equitable, resilient, sustainable, livable and safe cities

15.7 Recommendations

Before African cities integrate AI into their planning approaches, systems, strategies and processes, these core city-planning principles are recommended to achieve vibrancy, livability, resilience, inclusivity, innovation and smartness for their cities.

Design solutions that positively impact urban life: ICT-driven planning and solution design must align with and seamlessly integrate into the overarching urban development framework. This city blueprint, in turn, should articulate a vision that harmonizes spatial and technological elements through meticulous planning of both physical and digital infrastructures. Such a vision is inherently shaped by the city's heritage, geography, economy and collective aspirations.

However, at its core, the plan must prioritize the human dimension—citizens, service providers and investors. Urban planning must ensure that the devised solutions yield tangible benefits for the populace, enhancing access, efficiency, livelihoods, quality of life, sociocultural vibrancy, community vitality, citizen services and governance.

Design technology to improve city performance: ICT design within the overall smart-city design must aid and improve the performance of city functions and provide benefits to every stakeholder. Using design and systems thinking frameworks, planners must focus on accruing multi-parameter benefits including economic, infrastructural, social, cultural or environmental gains. Using these frameworks, planners must also recognize the interdependencies of functions and systems in a city, which, in turn, create constraints or opportunities for design innovation. Planners must design open, shared, interconnected, transparent, secure and ethical systems and data structures.

Design systems for mass use in local environments: The design of ICT and non-ICT components within smart cities in India must be meticulously tailored to align with the unique Indian context. The physical, climatic and behavioral characteristics of the country are distinctly singular. Designs should be crafted to accommodate the diverse Indian topography, weather variations and usage patterns. Best practices derived from established and mature cities in North America, Europe, or the technologically advanced metropolises of East Asia may not be applicable within the African framework. Urban planners in Africa must amalgamate a repertoire of global and indigenous solutions, sourced from both reputable firms and innovative thinkers, to devise strategies that resonate with African city economy, market dynamics and prevailing conditions. The monetization potential per user or per service in Africa may be comparatively lower than in global markets; thus, it is imperative to explore novel financial and investment frameworks. The burgeoning population in our cities exerts considerable pressure on resource utilization, necessitating that solutions be thoughtfully designed and personalized for millions of users.

Design solutions that can be built to last: Cities undergo development, evolution, and rejuvenation over extended time frames. Forward-thinking frameworks empower urban planners to conceptualize, design, and implement solutions that are enduring and resilient. Engaging with citizens and stakeholders is paramount to grasp their current, latent, and prospective expectations, desires, and motivations. Solutions must satisfactorily address these collective needs for the forthcoming decade or two. Technological advancements and architectural frameworks evolve with remarkable rapidity. Nevertheless, the urban and information and communication technology (ICT) designs envisioned and executed must remain functional, pertinent and operational over an extended duration.

15.8 Conclusion

The only way for African cities to meet the needs of the current times, and create smart cities through the city planners, urban managers, city governments and city leadership is to integrate AI into their city planning and management approaches, systems, processes and strategies. Therefore, in this 4th industrial revolution era of technological advancement, cities need to be much more innovative and must embrace technology as a key pillar in solving city challenges and city problems. The following list summarizes AI's potential for addressing urban challenges and problems and highlights its value for integration into city planning and management.

Data analysis and management: AI possesses the capability to analyze extensive volumes of data derived from a myriad sources, including sensors, satellite imagery, social media platforms and public records. It assists urban planners in discerning intricate patterns, emerging trends and significant correlations within metropolitan environments.

Predictions: AI algorithms possess the remarkable capability to forecast prospective scenarios by analyzing historical data, thereby enabling planners to anticipate trends in demographic expansion, traffic dynamics, infrastructure utilization and environmental transformations.

Decision optimization: AI optimization algorithms can assist in identifying the most effective solutions to intricate challenges, such as enhancing public transportation routes, minimizing energy consumption, or maximizing the allocation of green spaces within an urban environment.

Simulation and modeling: AI-driven simulations possess the capability to forecast the ramifications of proposed urban developments or policy alterations on various dimensions, such as traffic dynamics, air quality and communal well-being. This empowers urban planners to meticulously assess diverse scenarios prior to their enactment.

Smart infrastructure: AI facilitates the advancement of sophisticated infrastructure, encompassing intelligent transportation systems, energy grids and water management solutions. These systems possess the capacity to adapt in real time to evolving conditions, thereby enhancing both efficiency and sustainability.

Public engagement and decision making: AI tools can significantly augment public engagement in the planning process by offering interactive platforms for citizens to articulate feedback, visualize proposed initiatives and contribute innovative ideas. Moreover, natural language processing algorithms can adeptly analyze public sentiment derived from social media and online forums.

Risk management and resilience: AI possesses the capability to discern vulnerabilities within urban systems and devise strategies to mitigate risks associated with natural disasters, climate change, and other potential threats. This encompasses the development of predictive models for flood risk assessment, earthquake preparedness and comprehensive emergency response planning.

Fig. 15.4 Use of AI to create sustainable and self-reliant landscapes. *Source* Open AI (2024) Gemini generated image

Urban design and architecture: AI can assist architects and urban designers in generating a plethora of design alternatives, optimizing spatial configurations and integrating sustainable practices. Generative design algorithms possess the capability to explore an extensive array of possibilities predicated on delineated criteria and constraints.

Accessibility and equity: AI can assist planners in addressing challenges related to accessibility and equity by meticulously analyzing spatial data to pinpoint under-served communities, evaluate transportation alternatives and allocate resources in a more equitable manner.

Overall, AI enables urban and regional planners to make more informed decisions, optimize resource allocation and create sustainable, livable environments for current and future generations. However, it's essential to ensure that AI technologies are deployed ethically and inclusively to avoid exacerbating existing inequalities.

Figure 15.4 shows how AI can be utilized to create sustainable and self-reliant landscapes.

References

Alanda AK, Radoine H, Nyasulu C (2024) Urban challenges and strategies in African cities: a systematic literature review

Batty M, Axhausen KW, Giannotti F, Pozdnoukhov A, Bazzani A, Wachowicz M, Ouzounis G, Portugali Y (2012) Smart cities of the future. Europ Phys J Spec Top 214(1):481–518. https://doi.org/10.1140/epjst/e2012-01703-y

He Y, Zhou Y, Wu S (2020) Application of machine learning algorithms in optimizing urban spatial layout: a case study of Shanghai, China. Sustain Cities Soc 53:101944. https://doi.org/10.1016/j.scs.2019.101944Journal

Kadiri WA (2012) Urban growth and expansion of intermediate cities in Africa: planning response. Afric Plan Assoc (unpublished)

Kihato M, Kururi-Sebina G (n.d) Urban socio-spatial change and sustainable development: the neo-city phenomenon. Perspectives. In: Political analysis and commentary from Africa. Heinrich Böll Stiftung Foundation, Africa

Garth M (2010) Seven themes in African urban dynamics

UNDP (2012) Human development report 2012

UN-Habitat (2011) Cities and climate change: global report on human settlements

Watson V (2009) Seeing from the South: refocusing urban planning on the globe's central issues. Urban Stud 46:2259–2275. https://doi.org/10.1177/0042098009342598

Part VII
Future of AI Just Transformation of Cities: AI City Climate Planning Research and Applications

Chapter 16
Scientific Insights for AI-Based Urbanism Towards an Architect of City Climate Planning: A Scientific Assessment

Kwi-Gon Kim

Abstract City climate planning is a dynamic and still-developing new field. The scientific validity of the AI technology available for the prediction and optimization of climate solutions varies depending upon the particular climate actions. The primary purpose of this chapter is to emphasize how new AI technologies can improve planning processes and outcomes and promote inclusive and sustainable developments. More specifically, this chapter presents the diversity of innovative ideas on the state-of–the-art in this new area of interest of AI planning, particularly for inclusive participation and sustainable urban futures. It also briefly describes the results of the World Bank Planner Certification Program Workshop.

Keywords AI optimism · Synthesis science · AI agent system science · Urban computing science · AI scientific revolution · AI city solutions · Scientific insight · Planning domain · Digital marketing

16.1 Academic Perspective of the Demands of AI City Solutions

The future is built on optimism. AI optimism may fuel transformative technical and scientific abilities for city climate planning and development and reshape future cities through AI scientific revolution.

Scientifically speaking, AI city climate planning is a developing new field based on diverse disciplines such as synthesis science, citizen science, data and AI science, digital planning science, sustainability science, AI agent system science and urban computing science.

New scientific research and insights are needed to:

- Learn how we are staying on course for climate neutrality and resilience

K.-G. Kim (✉)
Seoul National University, Seoul, South Korea
e-mail: kwigon@snu.ac.kr

K.-G. Kim (ed.), *AI Transformation of Cities for Climate Solutions*,
https://doi.org/10.1007/978-3-032-21940-4_16

- Discover how climate action fuels innovation, jobs and growth
- Explore real-world success stories and future visions
- Engage with policymakers, cross-sector leaders and citizens.

The effectiveness, efficiency and practicality of development planning have increasingly been challenged by the complex nexus between physical, spatial, ecological, social and climate change problems associated with sustainability, the conflicting interests of stakeholders and citizen activism.

The scientific application of AI in sustainable development planning demands new ways of planning theory and proper frameworks and methodologies as well as tools and adjusted communication strategies between the parties involved, such as citizen science, as described in the previous chapters.

In this, there are planning practice needs from academia regarding the conceptual definition and methodological approaches of AI-based planning. Academics need a clear perspective of the demands of practice to develop more effective and implementable solutions.

The integrative literature review critically explores the use of AI in expanding smart cities across France, focusing on urban efficiency, sustainability, and safety. The French study examines the multiple challenges that French towns face when integrating AI technology to become smart cities, concentrating on technological integration, economic limits, data privacy and security, the digital divide, legal and ethical considerations and public acceptance. This problem impacts urban residents, as it influences their quality of life, access to services and environmental sustainability, necessitating a balanced approach to technology implementation that considers both benefits and potential social disparities. This study shows how France's smart city development can use AI to improve sustainability, urban planning, public safety, technological integration, and economic constraints (Ejjami 2024).

16.2 AI Planning Domains for Scientific Insights

AI Planning domains for scientific insights include urban regeneration, land use, energy transition and infrastructure, green–blue infrastructure, buildings, water management, waste management, NBS, green mobility, and smart lifestyle (e.g., food, clothing, fashion, education, etc.)

An action-oriented framework for extensive research and scientific methods is needed to integrate an analysis of urban climate factors, system efficiency, form, materials and green/blue infrastructure, assessing their effects on emissions, extreme events, justice and human comfort. Embedding climate-resilient development in planning, urban design and architecture ensures that AI transformation of cities advances carbon neutrality, adaptation and well-being for people and nature. These domains call on researchers and practitioners to bridge research and practice through developing consistent, multi-scale methods and tools that align outcomes with shared targets.

16.3 Innovative Ideas of the State-of-the-Art: Optimistic Process of Potential Realization

There are several prerequisites that must be fulfilled for a proper realization of innovative ideas of the state-of-the-art of AI transformation.

- Study of human–AI symbiosis to co-manage cities: More research on how planners, citizens, and AI systems can collaboratively make decisions
- AI carbon neutrality marketing: A key question is where and how to deploy AI Interdisciplinary teams: Interdisciplinary teams can provide decision makers with more complete information and enable better realization of a particular climate action
- Technology for AI/digital carbon neutrality at the level of city operation and citizen lifecycle. It is necessary to plan a coordinated, integrated approach to digital marketing, which is integrated with other communication activities (Chaffey and Chadwick 2019).

16.4 Certification of AI City Climate Planners

The continuing evolution of scientific assessment and practices of city climate planning process leads to the conclusion that there is a need for professional credentials for city climate planning. For example, the World Bank City Climate Planner Certification Program Workshop (2014) undertook a feasibility study for establishing a certification program.

The purpose of the workshop was to explore what role (if any) a professional certification program might play in advancing city climate planning efforts around the world. The key results from survey of potential target group for certification program revealed that:

- There is a need for some type of city climate planner professional certification program
- The proposed program would ensure the professionals are well-trained and knowledgeable in city climate planning process(es).

The author has co-hosted the workshop as Secretary General of Urban Environment Accords (UEA), and dedicated to close collaboration and strong partnerships (Photo 16.1).

Photo 16.1 Meeting of City Climate Planner Certification Program Workshop with 15 international agencies including UEA, World Bank, IDB, US Green Building Council, Global Envirnment Facility, LCLEI, UN ESCAP, World Resources Institute, UN Habitat, and Institute for Sustainable Development

References

Chaffy D, Chadwick FL (2019) Digital marketing: strategy, implementation and practice, pearson, Harlow England, p xiii

Ejjami R (2024) AI-driven smart cities in France. Int J Multidisc Res 6(3). https://doi.org/10.36948/ijfmr.2024v06j03.21920

GPT (2025) Website

World Bank (2014) City climate planner certification program-inception

Workshop Report (2014) Gwangju, South Korea, p 15

Chapter 17
Examples of Scientific and Practical Questions and AI Research and Development Topics

Kwi-Gon Kim

Abstract The process of transforming cities to AI has led to the delineation of scientific and practical questions and new research needs. This chapter presents some specific questions to be answered and potential AI research and development topics.

Keywords Research needs · Challenge and risk · AI education · AI workstation · Social network analysis · AI city networking · Flood defense design · AI technical transfer · Decentralized AI data center · AI function · Transformation pace, Green low-carbon AI life

17.1 Scientific Questions to Be Answered on AI Technology Research and Applications

The following list contains questions that represent opportunities and challenges associated with AI technology research and applications:

- Who are the AI tech visionaries working with city climate planners?
- To what extent can AI technologies improve planning and development processes and outcomes regarding effectiveness, efficiency and equity in terms of sustainable development?
- To what extent can AI technologies increase citizen engagement and reshape power relations in planning practice?
- Under what conditions can AI technologies add value to planning practice and promote sustainable development? What are the new challenges and risks?
- What kind of new data are generated by AI platforms and how can these be fed efficiently and effectively into planning processes?

K.-G. Kim (✉)
Seoul National University, Seoul, South Korea
e-mail: kwigon@snu.ac.kr

K.-G. Kim (ed.), *AI Transformation of Cities for Climate Solutions*,
https://doi.org/10.1007/978-3-032-21940-4_17

- What are the new methodologies for collecting, integrating, modeling and simulating multiple data sources to support complex decision-making processes in AI planning?
- What are the best practices of AI planning? What new insights can we learn from it for a more sustainable urban future?
- What are the consequences of planning education and curricular development, incorporating new knowledge, framework, methods, and tools?
- Which choices matter most for the climate?
- What could be the AI city markets?

In summary, we need to answer what contributions AI-based, climate-smart and sustainable eco-cities could make to industry 5.0 or business 5.0, emphasizing the collaboration between humans and machines to bridge city climate action with national climate plans and implementation of the global SDGs and Paris Agreement, including natural climate solutions.

17.2 Examples of AI Research and Development Topics Open to All Interested Scholars and Practitioners

Future research and development topics for AI-generated methods and techniques for climate solutions include the following:

(1) AI infrastructure (hardware): AI workstation.
(2) Integration of practice and theory-driven approaches.
(3) AI software.
(4) AI-Powered personalized services integrating into our daily lives such as healthcare, life style, food, clothing, access, smart home and green mobility.
(5) Applications of prediction and optimization for decision-making by using AI models:

    - Global and urban sustainability indicators
    - Urban modeling AI
    - AI-powered LCA and consumption/needs-based GHG emission accounting for climate action: Downscaled climate modeling for consumer behaviours
    - Geospatial artificial intelligence (GeoAI): It could provide a carefully optimized set of virtual urban form and spaces intended to meet climate-smart and sustainable growth indicators.

        - Land use AI: AI scenario-based, land-use modeling
        - Urban form (shape of cities) AI
        - Water management AI
        - Energy AI
        - Green-blue AI
        - Smart building AI

  - Green mobility AI
  - Remote sensing for geospatial science.

(6) Applications of location big data and spatial platforms:

- Human spatial behaviour, activity and mobility
- Place modeling and understanding of future land-use change simulation and scenarios
- Transportation and traffic system
- Social network analysis
- AI city networking for climate solutions.

(7) Social and ethical issues:

- Data quality and representativeness
- Privacy and ethical issues
- Social inequality across cyber and physical scope (digital divide)
- Governance and citizen science
- Race to regulate AI.

(8) Applications of AI in local participation:

- Generation of local and indigenous knowledge
- Data collection
- Master plan preparation
- Monitoring and evaluation
- Collective decision-making.

(9) Creation of AI Agents.

- Creation of customized multi-AI agents
- Creation of customized single-AI agents.

(10) Virtual cities and virtual ecosystems versus real cities and real ecosystems.
(11) Flood defense design: AI-powered methods, such as reinforcement learning, can design dynamic seawalls that adjust their height over time based on projected sea-level rise, making defenses more efficient and cost-effective than traditional static designs.
(12) Localized AI applications and technical transfer.
(13) Political and institutional frameworks to promote AI.
(14) Data-driven climate education powered by AI.
(15) AI-assisted analysis of ocean–atmosphere interactions and teleconnection with a focus on early warning system and integrated coastal zone management (ICZM).
(16) Decentralized data center aligning with decentralized energy supply chain.
(17) Advancement of AI function through AI training for climate solutions.

17.3 Conclusion

The future of AI transformation of cities for climate solutions will be characterized by manifold activities. China has implemented a joint action plan for 'coordination of digitalization and greening'. The key actions cover the green transformation of data centers, communication base stations, transport, ecological construction and urban operation, among others, which have effectively promoted the transformation pace of digital industrialization and industrial digitalization promoting the green and low-carbon transformation of economy (The China Internet Development Foundation 2022). Cities consume more than 70 percent of the energy, and urban energy consumption can be effectively reduced through urban AI transformation towards green and low-carbon AI life. The author views AI models proposed in this book as cooperative efforts towards innovative climate solutions.

Reference

The China Internet Development Foundation (2022) Digital carbon neutrality, p 26

Correction to: AI Transformation of Cities for Climate Solutions

Kwi-Gon Kim

Correction to:
K.-G. Kim (ed.), *AI Transformation of Cities for Climate Solutions*, https://doi.org/10.1007/978-3-032-21940-4

The original version of the book was inadvertently published with incorrect author names and affiliation in chapters 5, 6 and 14. The correction chapters and the book have been updated with these changes.

The updated version of this book can be found at
https://doi.org/10.1007/978-3-032-21940-4

K.-G. Kim (ed.), *AI Transformation of Cities for Climate Solutions*,
https://doi.org/10.1007/978-3-032-21940-4_18

Appendix
A Guide for AI–Based Ecosystem Approach to Minimize Negative Impacts of Sea-Level Rise in Coastal Cities

Kwi-Gon Kim

A.1 Introduction

There is growing interest in re-naturalizing coastal environments by adopting AI nature-based solutions (NBS) as a means to foster more resilient coastal cities.

A.1.1 Terminology of Sheltered Coasts

Mitigating erosion on sheltered coasts involves balancing property protection with ecosystem health, often contrasting 'hard' structures (seawalls) that increase erosion nearby and harm habitats with 'soft' nature-based solutions (living shorelines, vegetation, beach nourishment) that maintain natural features but may require more upkeep, while the original ecosystem services (habitat, water filtration, storm buffering, recreation) are often lost or traded for new ones (e.g., substrate for new organisms) with hardening, necessitating integrated planning for sustainable outcomes.

A.1.2 Guide Organization

This document provides a guide to reorienting shoreline erosion mitigation towards an innovative framework, and methods and techniques of mitigation to minimize the negative impacts of sea-level rise through a process with nine steps for the era of AI. Some of the stages are illustrated by tables or examples of current activity that can act as models to shape future mitigation practice.

K.-G. Kim (✉)
Seoul National University, Seoul, South Korea
e-mail: kwigon@snu.ac.kr

K.-G. Kim (ed.), *AI Transformation of Cities for Climate Solutions*, https://doi.org/10.1007/978-3-032-21940-4

A.2 Shore Erosion Mitigation Policies

A.2.1 *Mitigating and Regulating Development Impacts*

Wetlands mitigation refers to any activity taken to avoid or minimize damage to wetlands, and to restore, enhance or create wetlands as well.

The challenge for regulators is to find ways to improve the overall quality of shore wetlands while allowing developers the opportunity to obtain a reasonable return on their investment.

A.2.2 *Global Collaboration*

Global collaboration contributing to shaping the future of coastal restoration includes:

- Shaping policy
- Advancing cutting-edge science
- Implementing nature-based solutions (NBS)
- Building community partnerships
- Shoreline tech transfer
- Reimagining how to communicate impact with AI means: Through virtual representation of real-world application.

A.2.3 *Shoreline-Related Law and Governance Issues*

As an example of shoreline-related law and governance, in 1970 the coastal program began in New Jersey, U.S.A., following enactment of the Wetlands Act, and was expanded in 1973 with enactment of the Coastal Area Facility Review Act (CAFRA).

A.3 AI-Based Climate Change and Disaster Risk Models

The AI approach can provide frameworks and models that could be adapted to address sheltered erosion problems within a city or regional context to develop sheltered coast erosion mitigation plans.

Table A.1 AI-based climate-change adaptation: early warning system AI agent model

The five pillars of strategic activities	AI applications
1. Data streams (detection, observation, monitoring, analysis, forecasting)	AI systems to process and integrate diverse data streams into real time, including meteorological, geological, social media and historical climate records, etc.
2. Pattern recognition	Machine learning algorithms to identify complex patterns and anomalies
3. Disaster risk knowledge and management	AI predictive models to forecast the potential impact and severity of an event
4. Warning dissemination and communication	AI systems to streamline the communication of warning to vulnerable populations, aiming for timely and efficient alerts
5. Preparedness and responses	AI systems helping to improve disaster preparedness and responses

A.3.1 Whole Shoreline Approach: A Shoreline Framework

AI technology can be applied for each step of early warning activities, as shown in Table A.1.

A.3.2 Sheltered Coasts Approach: A Site-Based Framework

A list of five mitigation processes and AI applications are contained in Table A.2.

A.4 Nine Steps to Develop Erosion Mitigation Plans

The nine steps required in development of erosion mitigation plans for sheltered coasts are presented in the following sections.

A.4.1 Objectives

Setting clear objectives is crucial to mitigating erosion with AI. It needs to be part of mitigation process. During this process, mitigation projects should aim to:

- Examine the impact of shoreline on sheltered coastal environments
- Identify conventional and alternative policies, strategies and techniques to minimize potential negative impacts of sea-level rise

Table A.2 AI-based processes for the mitigation of sheltered coast erosion

Mitigation process	Description	AI application
Understanding of physical setting	(1) Sedimentation and hydrodynamic factors	(2) Long-term geomorphic monitoring by satellite imagery, LIDAR, etc. • Erosion and sand movement path analysis by AI models • Salinity sensors
Site analysis		• Optimization of maintenance strategy through wave and sedimentation models, and AI analysis • Wave risk analysis • Use of drone image and structural damage • Predictive model of underground water salinity • Prediction of underground water level to control salt water forward movement
Design criteria	• Level of protection • Acceptability for property owner and for budget requirements	• 3D printing • Human–nature interaction models
Designing the overall system	Four options to address erosion: • No action • Planning vegetation • Hardening • Adding sand (beach nourishment) • Trapping (groins or breakwaters) Prediction and optimization of options	• Prediction and optimization of mitigation material, input timing and quantity by AI models (digital twins. machine learning, etc.)
Implementing an effective mitigation strategy	Sectoral implementation plan: • Land use • Ecosystem and vegetation • Engineering (hardening) • Trapping (or adding sand) • Composite systems	• AI sensor data • AI-based automatic control system

- Consider adjacent or nearby coastal resources from a wetland connectivity perspective
- Develop models to predict and optimize the evolution of mitigated sheltered coasts under various scenarios
- Explore possibilities of using AI technologies in the whole process of mitigating shore erosion and inundation along sheltered coasts.

A.4.2 Selection of Sheltered Coasts

The second step involves identifying and mapping the shore erosion areas, within the boundary of a project, to illustrate the mitigation framework, methods and technologies.

A.4.3 Description of the Selected Site: Understanding of Physical Setting and Sea-Level Rise

This step involves assembling information on the existing physical setting and sea-level rise in the area of the mitigation project, particularly focusing on sea-water quantity, tide and sedimentation parameters related to anticipated sea-level rise. Sources of information include relevant city, county, province and federal agencies that have monitoring programs.

For example:

- Global average sea-level has risen 8–9 inches (21–24 cms) since 1880 (Google website 2026).
- The increasing height of the Korean ocean is 11.5 cm during the period from 1989 to 2024 (36 years). Annual sea level has risen around 2.6–3.6 mm (Korea National Ocean Research Institute (NORI) 2026).

The following photos illustrate examples of existing coastal areas (Photos A.1, A.2, A.3, A.4 and A.5).

Photo A.1 Mangrove forest preserve, Port Moresby, Papua New Guinea (PNG). *Source* Photo by the author

Photo A.2 Sand dune preserve, Jeju Island, Korea. *Source* Photo by the author

Photo A.3 Urban devlopment along coastal area, Sokcho, Korea. *Source* Photo by the author

A.4.4 Mitigation Technologies Available

A.4.4.1 Shoreline Hardening Technology

Cumulative effects of hardening should be considered in shoreline mitigation plans, both values invested by the affected communities in non-hardened shorelines and the value of ecosystem properties that stand to be lost with shoreline hardening.

Photo A.4 Condominium development encroaching coastal salt marsh, Yangyang, Korea. *Source* Photo by the author

Photo A.5 Tidal wetland offering a vital resting and feeding place to migrating waders and other shorebirds. *Source* Google website 2026

Proactive erosion mitigations should be implemented to avoid unintended consequences when hardened shorelines reduce the recreational, aesthetic, socio-economic and ecological value of sheltered coastal areas.

Examples of hardening practices are shown in Photos A.6, A.7, A.8 and A.9.

Photo A.6 Stone groin: natural stone hardening (Jeju Island)

Photo A.7 Natural stone wall revetment (Jeju Island)

A.4.4.2 Vegetation Technology: Strips of Wetlands to Control Erosion of Shoreline

The construction of wetlands has proven effective at dissipating wave energies and preventing further erosion and undercutting of the shoreline. Notably, the cost of construction wetlands compares favorably with more traditional erosion control methods, such as stone revetment or wooden bulkheads. The typical approach is outlined as follows:

(1) Clearing the shoreline: First the shoreline is cleared to ensure that the plants will have access to full sunlight.
(2) Planting wetland grasses: Next the site is planted with nursery-grown grasses (e.g. salt-marsh hay) considering the salinity of the water.

Photo A.8 A composite system of natural stone revetment (left) and concrete sea wall (right) (Jeju Island)

Photo A.9 A dune beach erosion: Fencing and wetland vegetation may help to induce baffling and sand settlement (Jeju Island)

(3) Filling with sand and regrading: The site is then restored to its former (pre-erosion) elevation by filling with sand and regrading.
(4) Establishing the marsh: After several months, the marsh is well established.

Examples of ecosystem-based practices are shown in Photos A.10 and A.11.

A.4.4.3 Managing Land Use

Land use needs to be managed through planning, regulation, incentive and acquisition. Some examples of regulated land use are shown in Photos A.12 and A.13.

Photo A.10 Protected natural coastal ecosystem without bulkheads: salt marsh grass (Jeju Island)

Photo A.11 Protected natural coastal ecosystem: a combined ecosystem of forest, salt marsh, natural stone and sand beach (Jeju Island)

A.4.4.4 Summary of Ecosystem Services Assigned to Mitigation Techniques

Table A.3 provides an overview of the ecosystem services offered by natural coastal environments alongside those delivered by standard techniques employed to mitigate shoreline erosion.

Photo A.12 Unregulated land use with stone revetment: development encroaching coastal wetlands (Jeju Island)

Photo A.13 Land use policy and zoning regulation: landward placement of development leaving vegetation and beach intact at least for the near term (Jeju Island)

A.4.5 AI-Driven Technologies: Integrating AI Technologies into Erosion Mitigation Strategies

(1) Prediction and optimization of future change with digital twin and machine learning: What will happen if nothing is done?

- The erosion process will most likely continue into the near future unless there is historical evidence of impending change in patterns of shore evolution

(2) Digital twin (virtual twin) simulation

Table A.3 Summary of ecosystem services provided by natural coastal ecosystems as well as by commonly used techniques to mitigate shoreline erosion

Ecosystem Services	Natural Coastal Ecosystems							Techniques to Mitigate Shoreline Erosion					
								Harden shoreline		Trap or add sand			Plant it
	Sandy beaches	Sand dunes	Mudflats (microalgae)	Marshes and mangroves	Seagrasses and macroalgae	Bluffs		Bulkheads and seawalls	Revetments	Groins	Breakwaters and sills	Beach nourishment	Marsh/mangrove planting
Habitat *fishes*	○	○	**○**	●	●	○		○	**○**	○	●	**○**	●
mollusks	●	○	●	●	●	○		**○**	**○**	**○**	**○**	**○**	●
crustaceans	**○**	○	**○**	●	●	○		–	○	○	○	○	●
turtles	●	**○**	–	○	**○**	○		–	–	–	–		○
birds	**○**	●	●	●	○	●		–				**○**	●
Nutrient uptake/cycle	**○**	○	**○**	●	●	○		○	○	○	○	○	●
Food production	○	○	**○**	●	●	**○**		○	○	○	○	○	●
Wave attenuation	**○**	●	○	**○**	**○**	●		○	**○**	**○**	●	○	**○**
Sediment stabilization	–	●	**○**	●	●	–				**○**	●		●
Gas regulation	○	–	○	●	●	–		–	–	–	–	–	●
Biodiversity	**○**	●	**○**	●	●	●		○	○	○	○	○	●
Recreation	●	●	○	●	●	**○**		**○**	○	**○**	**○**	●	●
Raw materials	●	●	○	●	●	●		○	○	○	○	○	●
Aesthetic value	●	●	**○**	●	●	●		–	–	–	–	●	●

Source National Research Council of the National Academies (2007) Mitigating Shore Erosion along Sheltered Coasts, the National Academies Press, Washington, D.C., p. 93

- What if vegetation is planted on the shoreline?
- What will be the consequences of hardening the shoreline?
- What if sand is added (beach nourishment) or trapped (groins or breakwaters)?
- Integrating AI into mitigation design process.

A.4.6 Design Elements and Criteria of Mitigation Practices

A.4.6.1 Design Elements of Sheltered Coasts

- Unconsolidated upland bluffs

- Dunes
- Beaches
- Intertidal (e.g., marsh, mangroves) vegetation
- Subtidal (e.g., macroalgae, seagrasses) vegetation
- Tidal flats
- Sandbars.

A.4.6.2 Design Criteria

- To test and verify reliability and effectiveness of selected techniques considering a possible acceleration of relative sea-level rise with AI
- To improve wetland connectivity
- To mimic nature including ecological art
- To consider practicality of proposed mitigation technologies
- To create ecosystem goods and services including biodiversity.

A.4.7 Mitigation Methods and Design for the Selected Sheltered Coasts

A.4.7.1 Erosion Mitigation Methods

- Engineering techniques
- Land management planning measures
- Ecosystem-based measures
- Mix of technologies.

A.4.7.2 Design Process and AI Applications: A Design Framework

See Table A.4.

A.4.7.3 AI-Driven Innovative Mitigation Methods and Design Details: NBS Innovation in Coastal Protection

- Selection of natural coastal ecosystems as reference sites
- Collection of monitoring data with AI sensors
- Design of prototypes of shore erosion mitigation in bluff and salt marsh, for example (Fig. A.1).

Table A.4 Technical innovations for wetland restoration

Component	Smart technology	
1. Data collection and analysis for preparation of wetland inventory maps	• Satellite image • Drones as airborne robots	• IoT/blockchain • TMS
2. Wetland assessment	• Cloud computing • Big data & analytics	
3. Wetland planning	• GPS • GIS	• Off-grid system • On-grid system
4. Wetland restoration design	• Virtual visualization • Connected-tech	
5. Construction oversight	• Digitalized work process	
6. Environmental monitoring	• IoT • Drone • Big data & analytics	• Cloud • Computing AI
7. Evaluation and feedback	• IoT open data • Accumulation of infra-test knowledge	
8. Community outreach	• Access to public open data: multi-directional • Blockchain & server	

A.4.8 Monitoring Data

The following data monitoring is needed to document the effectiveness of the employed techniques used with AI technologies:

- Selection of post-construction sites
- Monitoring for expected lifetime of the restored ecosystems
- Evaluation of the ecological consequences of alternative shoreline stabilization measures.

A.4.9 Implementation Plan

This phase concentrates on implementation plans. The ultimate outcome of an implementation plan should not be the plan itself, but institutional reform followed by mitigation action. Monitoring should be realized through internal auditing and reform of procedure, rules and standards.

4.9.1 The Mix of Technologies to Be Employed
4.9.2 Land Use Plans to Be Implemented
4.9.3 Ecosystem-Based Mitigation Projects to Be Undertaken

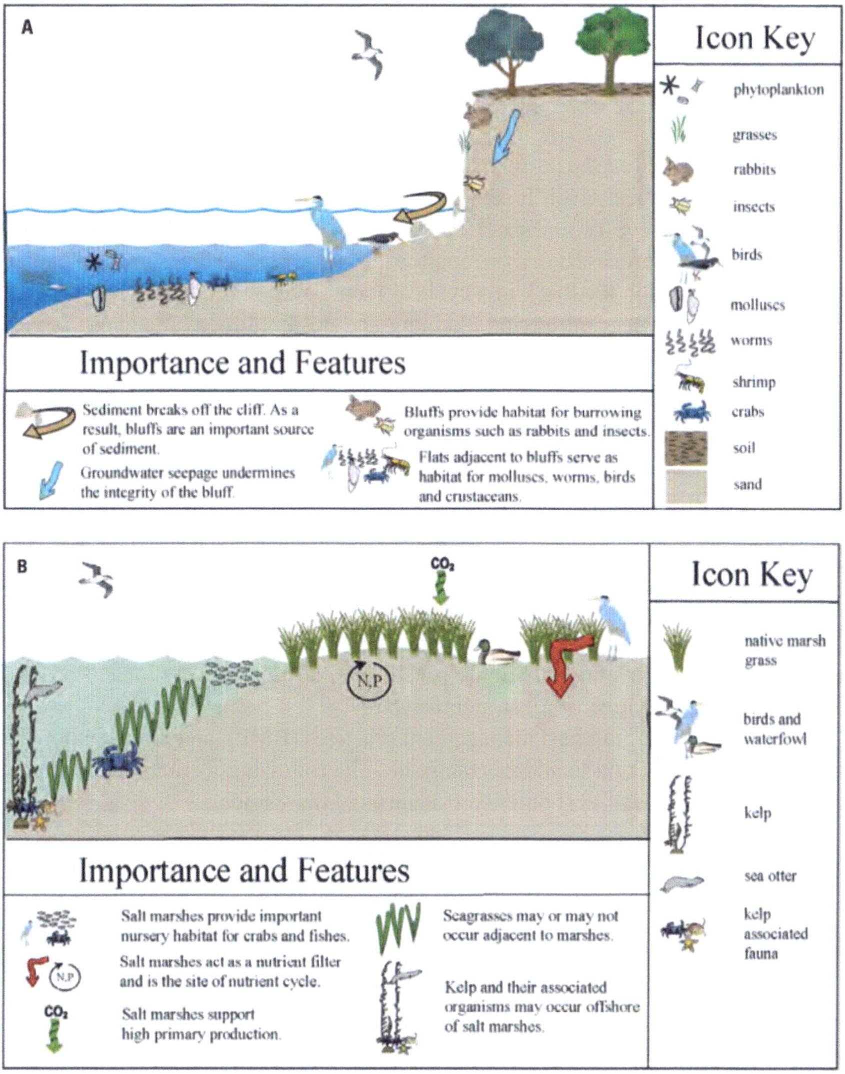

Fig. A.1 Conceptual diagrams of ecosystem service provided by a bluff (A) and a salt marsh (B) illustrating some typical components of the ecological communities and outlining the process that characterize these coastal environments. *Source* National Research Council (2007)

4.9.4 AI Platform Workflow to Be Integrated
4.9.5 Shoreline Erosion Policies and Design Criteria to Be Evaluated

A.5 Conclusion and Recommendations

A.5.1 Conclusion

This document reviewed the current state of, and trends in, shore erosion mitigation at the policy and project level. It set out the stages of shore erosion mitigation that are necessary to achieve a coordinated approach. In essence, the process is cyclical; a continuous learning process based on monitoring data, where AI helps in coordination and in taking action on feedback in a cyclical loop.

AI applications affect mitigation planning and design directly, and so it is vital to gain support and to encourage authorities, local people and businesses to change practices. The mitigation plan should weave the policy and planning threads together into a coherent multi-agency plan. To do this, it is necessary to understand the AI integrated process, and develop the skills of collaborative working.

A.5.2 Recommendations

The experience of shore erosion mitigation across the world shows very clearly that moving from conventional simple practices to sustainable, ecological process adapted to climate change is not straightforward.

This requires emerging best management practices (BMP), and many authorities have begun to take action to address the issue. The following recommendations are drawn from examples of best practice from around the world.

(1) Adoption of a new shoreline management framework: AI sustainability cost-benefit analysis of spectrum of potential solutions to shoreline erosion problems can be used to change the current management framework.
(2) Assessment of cumulative impact of erosion mitigation projects with reference to potential acceleration of sea-level rise.
(3) Choices of sustainable and climate-responsive mitigation methods and techniques: available for NBS.
(4) AI transformation of shoreline mitigation in the age of climate change: AI can be used in the mitigation site survey, as well as in the analysis, prediction and optimization, planning and design processes.
(5) Introduction of new permitting systems. Decision making should be based on:
    - Physical or ecological information to evaluate ecological benefits.
    - Potential cumulative impacts to avoid parcel-by-parcel approaches.
    - Types of erosion-mitigation approaches and their effectiveness under various site conditions.
    - Applications of AI technology in the whole erosion-mitigation planning and design process: Responsible AI deployment.

- Types of natural disaster insurance: Transparent and objective calculation of the insurance fee.
- Local participation in mitigation decision making.

It is recommended that these practices be used to develop international guidelines for mitigating erosion on sheltered coasts, and to promote ecologically beneficial measures through international dialogue.

References

Baines C Smart J (1991) A guide to habitat creation. A London Ecology Unit Publication

IDB (2018) Disaster and Climate Change: Risk Assessment Methodology for IDB Projects, A technical reference document for IDB project teams

National Research Council of the National Academies (2007) Mitigating shore erosion along sheltered coasts. The National Academic Press, Washington, D.C.

Salvesen D (1991) Wetlands: mitigating and regulating development impacts, 2nd edn. The Urban Land Institute

Treweek J (1999) Ecological impact assessment. Blackwell Science

The manufacturer's authorised representative in the EU is Springer Nature Customer Service Centre GmbH, Europaplatz 3, 69115 Heidelberg, Germany. If you have any concerns regarding our products, please contact ProductSafety@springernature.com

Printed and bound by CPI Group (UK) Ltd, Croydon, CR0 4YY
21/09/2026
02219991-0010